D1211922

Electric and Gas Welding

by E. F. Lindsley

Drawings by Forest J. Battles

POPULAR SCIENCE BOOKS

This book explains how to use electric and gas welding equipment to join and cut metals. Throughout the book, the author gives detailed safety precautions that should be observed when handling this equipment. These precautions must be followed faithfully if the operator of welding equipment is to avoid accident. However, since the correct handling of the equipment is beyond the control of anyone but the operator, the author and publisher disclaim responsibility for injury to persons or property that may occur through its use.

Copyright © 1981 by E.F. Lindsley
Published by Grolier Book Clubs, Inc.

Brief quotations may be used in critical articles and reviews. For any other reproduction of the book, however, including electronic, mechanical, photocopying, recording or other means, written permission must be obtained from the publisher.

Library of Congress Catalog Card Number: 77-26479
ISBN: 0-06-090723-1

Third Printing, 1987

Manufactured in the United States of America

Contents

Introduction

With this book and a low-cost electric arc or gas welding outfit you can learn the exciting and satisfying fun of welding within an hour. Your welds will be crude and won't begin to pass the tests for professional work. But you'll be able to weld together angle iron, steel plates, pipe, and tubing. Or, if artifacts are your interest, you'll find it easy to build metal art objects. You'll also be able to torch-cut or arc-cut steel and to braze or hard-solder such materials as cast iron, brass, copper, and light pieces unsuited for welding.

Many new electric and gas welding rigs have come on the market recently. Their cost is low enough to make them totally practical for home shop use. Many electric welders simply plug into a household outlet. Some gas torches are so small that you can weld tiny parts on the kitchen table. And there are cutting outfits you can lug with one hand. Best of all, they're all fun to use. This book will help you choose what you need.

Welding is one of the easiest arts to learn and one of the most difficult to master. The old-time blacksmith skillfully forge-welded wagon-wheel rims, hardware, pothooks, and horseshoes. But he also served a long apprenticeship. Today, the smith's forge has given way to specialized, often exotic, industrial welding methods. Both the blacksmiths and industrial welders were and are masters. This book does not pretend to prepare you for advanced industrial welding. If you want to acquire the skills to use sophisticated welding machines, do pipeline or pressure-vessel welding, or even hand-forge horseshoes, your best choice is a training school. But if you want to be able to do ordinary welding jobs around the home, and enjoy doing them, read on.

—EFL

1 Joining Metals by Welding

Simply stated, and with apologies to those who prefer more technical definitions, welding amounts to nothing more than melting two pieces of metal so they run together and become one. An easily understood example would be the common experience of having ice cubes stick together. Take a tray of ice cubes from the freezer and each cube is an individual piece of ice. Let them warm until they start to melt. Put them back in the freezer and the water-coated surfaces will merge and freeze together. You've "welded" ice. The only real difference is that ice melts at about 32° F. and steel melts at about 2700° F.

If this example seems oversimplified, it's not. There are some other lessons to be drawn from it. Note that each cube is formed from the same substance, water. There's no problem in making water mix with water. The material in our ice cubes is compatible. But if we use one ice cube and one plastic cube we won't get a very good joint. The substances are different; they melt at different temperatures, and they won't flow together. In short, they're not compatible. Nor, for example, are aluminum and steel, at least for ordinary welding techniques.

Let's carry our little example of the ice cubes a bit further. Suppose—I can't imagine why unless you're into ice sculpture—you want to join some cubes but the surfaces are a trifle rough and don't fit together very well. They'll only stick where the high points touch. No problem; you just dribble in a little water and fill the gaps. That's called filler material in welding. In arc welding the electrode itself (we'll explain this later) melts and runs into the gaps. In gas welding you hold a torch in one hand and a piece of filler rod in the other. You melt off enough filler rod as you go along to fill the gaps. Note that like the water we used with the ice cubes

Partially melted ice cubes join themselves together by fusing the water between their adjacent surfaces. When two pieces of metal are melted along adjoining edges so the metal fuses together, we call it welding.

the filler material must be of just about the same composition as the metal we're welding. Again, the compatibility factor. As an interesting sidelight, some arc welding electrodes are even designated as "fast-freeze" or "fill-freeze" type, so our ice-cube analogy is really quite accurate.

If you've ever watched a weldor at work—and I hope you used eye protection when you did—you may have been so distracted by the flaring of the arc or the torch flame that you didn't really see what was going on. When you get your own equipment and start practicing you'll soon find out that the actual area being heated to a molten state is quite small. In fact, it's seldom over ¼″ or maybe ⅜″ in diameter in spite of all the fussing and fireworks. This area of concentrated heat where the metal is molten and flowing like water is called the "puddle." I'll talk about the puddle

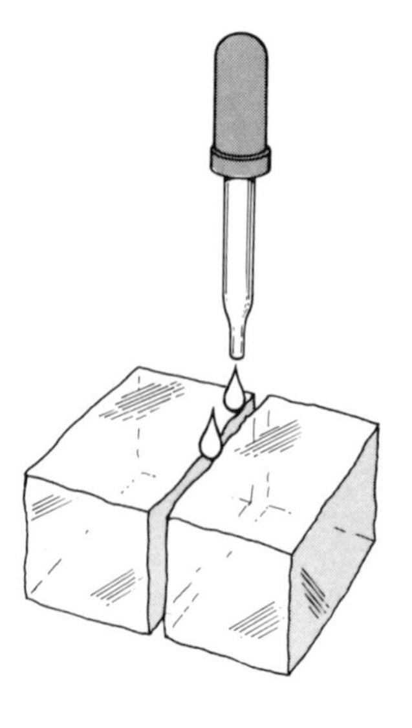

Even ice cubes wouldn't make a strong joint if only the high spots touched. Filling the gaps with water makes a complete bond. In welding we use molten metal from a "filler" rod to fill the gaps and build up the joint.

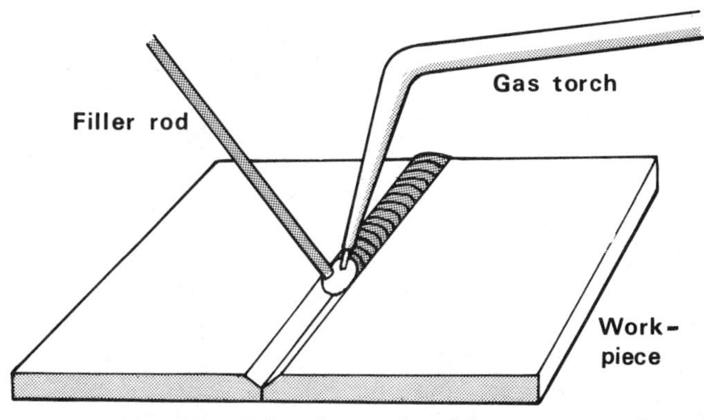

The gas weldor supplies filler rod to form a bead between two pieces heated to a liquid state by the torch flame. Filler rod and work metal must be compatible.

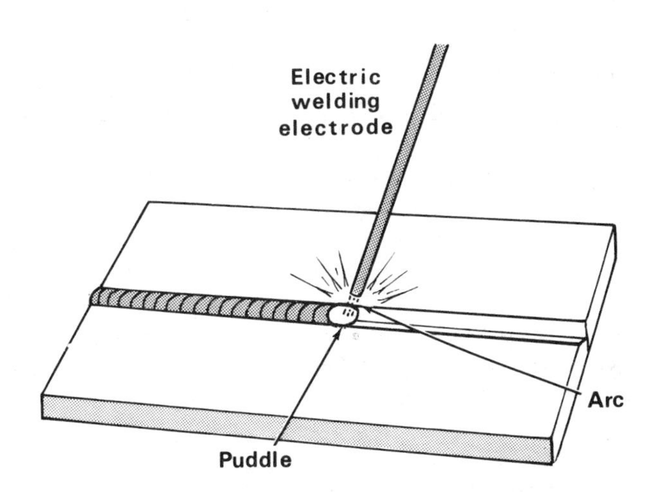

Very high temperature of electric arc between work and welding electrode melts workpiece to form a puddle and also melts electrode metal to provide filler. Puddle is confined to a small area.

frequently, because that's where the job is being done and without a puddle you don't get a weld.

The whole idea, then, of either gas or electric arc welding is to deliver enough heat to a concentrated area to melt the metal. Most ordinary steels used for home projects melt in the 2700–2800° F. range. If you've had experience using a blowtorch or a propane or MAPP gas torch, or even the latter assisted with oxygen, you know that although you can bring metal to a bright heat you cannot concentrate the heat or get it hot enough to form a good puddle. MAPP gas (stabilized methylacetylene and propadiene) is a recently introduced gas producing a flame almost 1000° hotter than propane but not as hot as acetylene and with different

Arc welding is better than gas welding for the thick channels and steel plates which require heavy welds in this home-built generator chassis. Gas welding would be much slower and cost more. The 230-amp transformer welder, right, is typical of home-shop welders now available at modest cost.

flame characteristics. Such heat sources can be used for other joining methods, such as brazing, silver soldering, and soft soldering, and they're handy for bending and forming metal, but they're not suitable for welding.

The only two practical heat sources for home welding are the electric arc or oxygen and acetylene gas. Very light welds, of a sort, can be made with MAPP gas and oxygen. The electric arc generates a temperature of 7000° to 9000° F. An oxyacetylene torch produces a flame of about 5700° F. Moreover, with both sources the heat is very concentrated so you can confine your welding area nicely and produce a neat, welded joint exactly where you want it. In a moment I'll explain the choices between the two heat sources for different types of jobs.

ARC WELDING. In the case of the electric arc, one cable comes from the "ground" connection of the welder and is clamped to the workpiece, thus grounding it. Another cable from the welder comes to a heavily insulated second clamp which you hold in your hand and in which you've secured a coated metal rod called an electrode. These two connections, the ground clamp and the electrode holder, plus your workpiece, make a complete electrical circuit from the welder and back to it. When welding you touch the electrode briefly to the work and then withdraw it slightly. The gap between the end of the electrode and the workpiece forms the intensely hot electrical arc.

Since the voltage involved is considerably lower than 120-volt household current, your chances of getting a shock are low—not so low, however, that safety precautions are not in order, and we'll get to that.

The tremendous heat of the arc described above soon melts a puddle in the workpiece and also melts the metal core of the electrode, which flows into and fills the puddle as you move the arc along the joint. Since the molten metal wants to combine with the oxygen in the air (which basically means it wants to burn), the coating on the rod is designed to melt and form a glasslike slag which floats on the metal and keeps the oxygen away. When the weld cools, you chip away the slag. You'll notice that I sometimes refer to the electrode as the "rod." That's because weldors do so in ordinary conversation. "Electrode" is more accurate, but "rod" is more commonly used.

GAS WELDING. Gas welding employs different equipment but is not different in principle from electric arc welding. Instead of electrical power, acetylene and oxygen from separate tanks are brought through pressure regulators and hoses to a torch. The two gases are mixed in the proper proportions in the torch. This mixture exits from the torch tip and burns with one of the hottest flames known.

In gas welding you direct this flame, an intensely hot cone, at the metal to form a puddle. This doesn't happen quite as fast as with the electric arc, but unless your workpiece is massive you'll see a puddle form in a few

Electric arc welder converts household power to lower voltage and higher amperage. Welder output is fed to electrode holder, across the arc, where it produces heat, and back to welder in complete circuit. Home-shop welders operate on 115 or 230 volts.

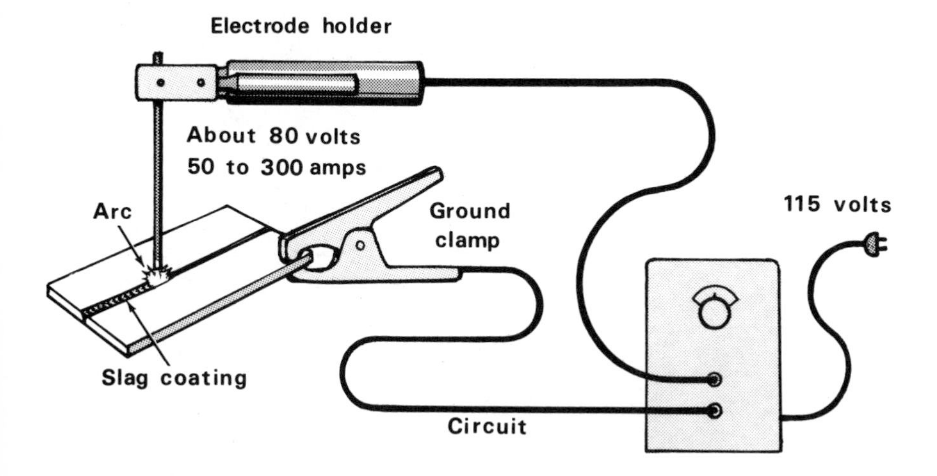

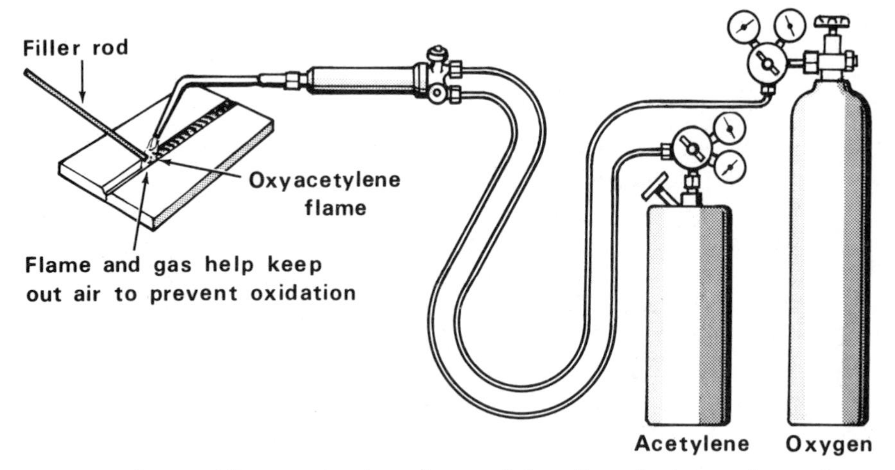

In oxyacetylene welding, combustion of a carefully adjusted mixture of acetylene gas and oxygen supplies heat to metal and filler. Fuel gases are stored under pressure; pressure is reduced and regulated before feeding gases to supply hoses.

seconds. The shield formed by the burned gases helps to exclude air and keep the metal from burning much as does the slag from the electric arc rod.

Filler rod is not always needed for gas welding. With two upfolded edges, for example, all you have to do is move the flame and puddle along and let the edges run together. Most of the time, however, you'll find it necessary to feed in filler rod to produce a bead or fillet. A fillet is the name given to filler metal in a corner that blends into each workpiece much like a cove molding between a wall and floor. One difference between this type of gas welding and arc welding is that both hands must work to-

Some welds, such as simple edge weld, require no filler rod and are made by simply heating metal so molten edges merge and fuse.

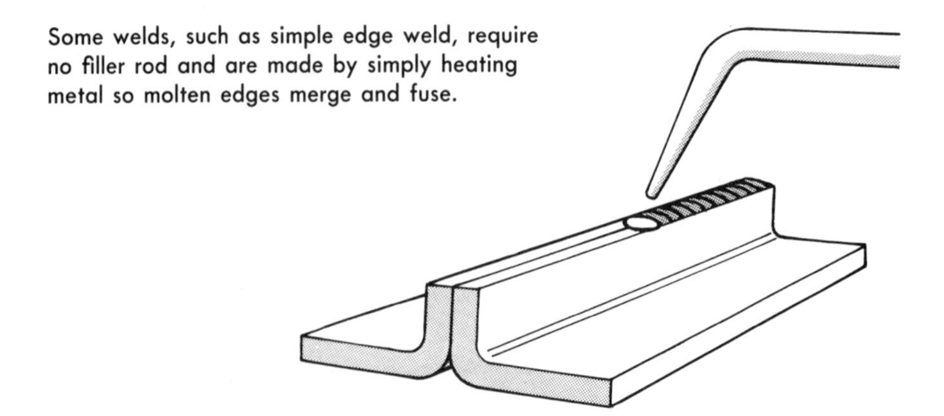

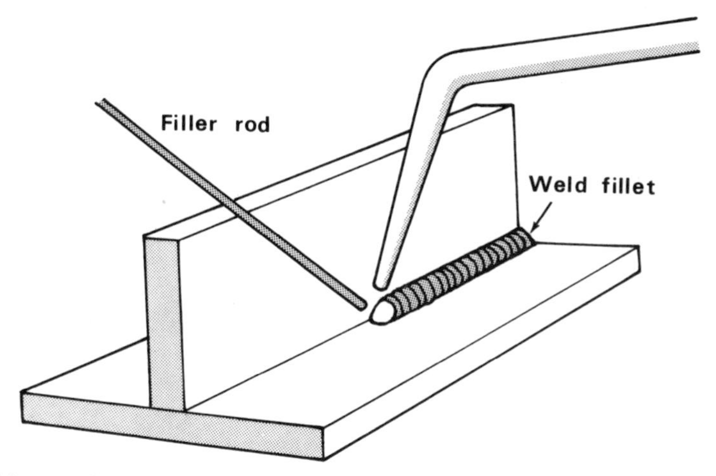

In welding, a fillet is a buildup of filler metal in the corner of a joint. Filler rod is almost always needed.

gether. One holds the torch and the other directs the filler rod. The rod is melted by dipping the end into the molten puddle.

So these are the basics of electric arc and gas welding. Now you need to know the types of equipment available and how to choose that equipment to suit your special wants. And you need to know how to get started using this equipment and how to weld safely and use the little tricks that make welding easier and keep you from winding up with a poor job. Let's look at the equipment first. Understanding it will make the process of learning to weld a lot easier.

Light, thin metal of theme table was suited to delicate touch of gas welding and brazing. Such materials heat and weld quickly, so cost of fuel gas is nominal.

2 Electric Arc Welders

Ideally you'd like to have both oxyacetylene and electric arc welding equipment in your shop. Why? After all, don't they do essentially the same job? The answer is yes, but they don't do all jobs equally well. If you're going to weld delicate art objects of very light metal, wire, and small pieces, the gas welder is the best choice. For tubular aircraft structures, race-car parts, and the like, gas is often the method of preference. But if your jobs may run from repairing or mounting a trailer hitch to building new support structures for a pier or fabricating a steel pipe swing set for the kids, or just general repair and construction, then arc welding is a better and less costly choice. I recommend a simple transformer arc welder for the beginner. Here are the reasons:

- Lower initial cost
- A single-cost purchase
- Trivial operating cost
- Better for common household jobs
- Easier-to-learn skill
- Less delicate and less critical equipment

Costs of Arc vs. Gas. Each of those reasons bears some discussion. The mail-order price of a 20-to-70-amp adjustable-range arc welder, complete with helmet, electrode holder, and carbon-arc torch, was $78 at the time of writing. A smaller 50-amp nonadjustable welder with the same accessories was only $58. That's it—complete, ready to plug in and weld. True, these light-duty units won't handle very big jobs or heavy metal and they're slightly harder to use than the heavier welders, but they're fine for

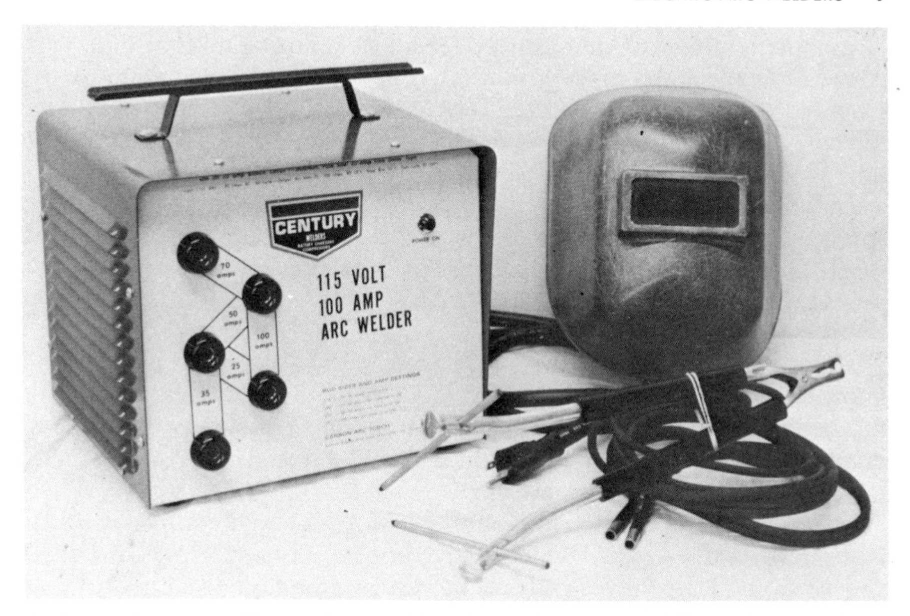

Package electric-arc kit, ready-to-weld, offers 100-amp capability, adequate for most home jobs, and requires only a standard 30-amp wall outlet. Combination electrode holder and carbon-arc torch, under helmet, works, but you'd probably want to upgrade it eventually.

many home shops. Even top-of-the-line AC/DC 230-amp units may cost less than $300.

In comparison, the least expensive real oxyacetylene outfit costs about $150. Unfortunately, when you've got it you're not ready to weld. Now you must go to the supply house, make a deposit on an oxygen and an acetylene tank, pay for the gases in them, and lug them home. After that, each month you'll get a bill for tank rental. How much depends on where you live and the cost of welding supplies there. An alternate option to renting tanks is a lifetime lease on your tanks, which will cost $200 or more depending upon locality. Naturally, you have to buy the gas and take the tanks in for exchange.

None of that applies to electric welding. The amount of power you use depends, of course, on how much welding you do. Unless you're in business doing production welding it's doubtful that you'll ever detect the power cost on your electric bill. You do have to buy electrodes, but unless you are working in exotic metals their cost is nominal.

Convenience. Aside from delicate jobs, you'll find that ordinary repair or construction welding on angle iron, steel plates, pipe, and such routine materials goes much faster with the arc. As soon as you strike the arc you're welding. With a gas torch a certain warm-up time is needed, and if the parts are fairly husky—a trailer hitch, for example—you'll burn a lot of gas just getting a puddle started. Moreover, during all that time heat is spreading through the workpieces. If there are gaskets, rubber or plastic parts, oil seals, or grease-packed bearings near the weld area you'll have to remove them or pack them with wet rags to try to keep the spreading heat from damaging them.

The electric arc is fast and there is no extended time for the heat to spread. Once the weld is completed and partially cool, you can cool the whole area down with water before the heat hits a vulnerable area. It's practical convenience such as this that makes arc welding so handy around the house.

The quality of any weld, gas or electric, depends on your skill and knowledge. My own experience has convinced me that it's just easier for the beginner to learn to strike an arc and hold it properly than it is to manipulate a torch in one hand and a filler rod in the other. Others may not agree; it's a matter of opinion.

The ruggedness and dependability of arc equipment compared to gas is unquestioned. Gas welders must have precision pressure regulators and gages with delicate valves and springs inside. Torches and torch tips cannot take abuse. If they are left unused for months, as often happens in the home shop, such components tend to get sticky, corroded, and balky. A transformer welder can sit unused on the corner of your workbench for a year, but plug it in and switch it on and it's ready, willing, and able.

Pressure regulators and gages for oxygen and acetylene welding gases have relatively delicate internal parts and must be treated with care in use and storage. Oxygen regulator, right, reduces pressure from over 2000 pounds to about 5. Acetylene regulator handles about 250 pounds. Both deserve respect from user.

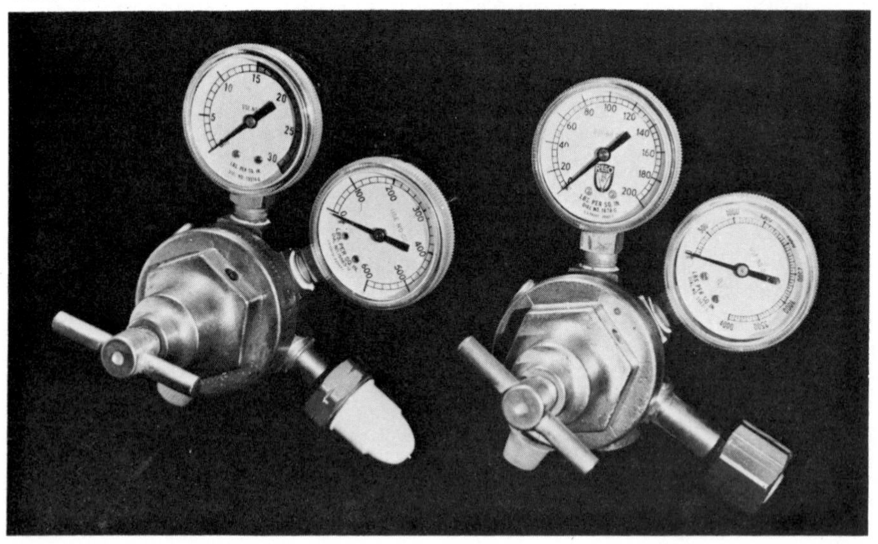

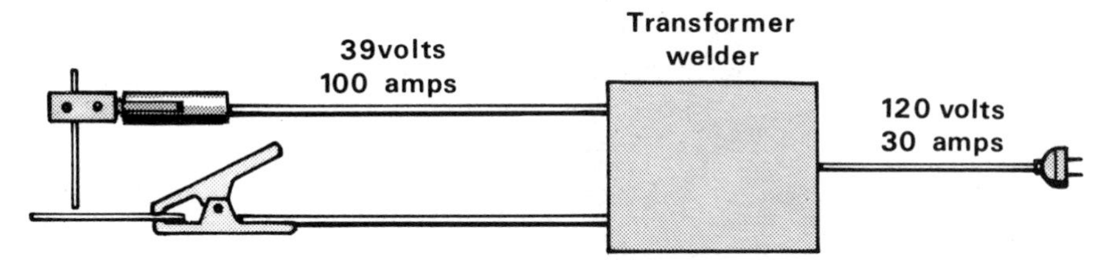

Welding transformer trades voltage for amperage. Newcomers to arc welding are often surprised that a 30-amp household circuit can provide 100 amps of welding current. The answer: 115-volt input is reduced to about 39 volts at workpiece.

TRANSFORMER WELDERS. The simplest kind of arc welder, and the most practical for home use, is the transformer type. Many beginners find the transformer concept confusing. They see displays or catalogues listing welders which will deliver 100, 200, or 250 amperes of welding current, and when they check their house wiring are dismayed to find that they have only, say, 100-amp service. Their question is perfectly logical—how can I take 200 amps out of a welder when my house service is only 100 amps or less? Or they say, "Here the catalogue states that I can plug a 100-amp welder into a 30-amp household circuit. That sounds impossible." It's not impossible and it works just fine. The secret is the remarkable ability of the transformer to step down voltage and step up amperage. It's a matter of trade-off and can work either way. For example, if the relatively modest boost in amperage with a welder sounds unreasonable, think of your car's ignition coil, which is also a transformer. It boosts the 12 volts from your car battery (actually only 6 volts with the ballast resistor in the circuit) to about 30,000 volts at the spark plug, but at very low amperage. A transformer welder, then, is a remarkable device which uses the principle of magnetic induction to reduce the voltage and step up the amperage. The result is the delivery of a large amount of electrical energy to a concentrated area of the workpiece to melt metal.

How your arc welder is made. If you like to take things apart to see how they work you'll be disappointed with a transformer welder. Take off the decorative and protective housing and there's nothing exciting to see. You'll find a metal core made up of plates of sheet iron stacked together. You'll also see a number of coils of wire wound on this stacked metal core. If the welder offers several amperage selections there may also be wires leading from various points on the coils to the taps where you plug in the cables, or there may also be a means of moving a section of the core to adjust amperage. But that's it—an iron core with a few coils of wire. You also will probably find a small motor and fan. That's to cool the coils. It hasn't got a thing to do with the actual welding any more than the fan on your car engine has to do with the drive line.

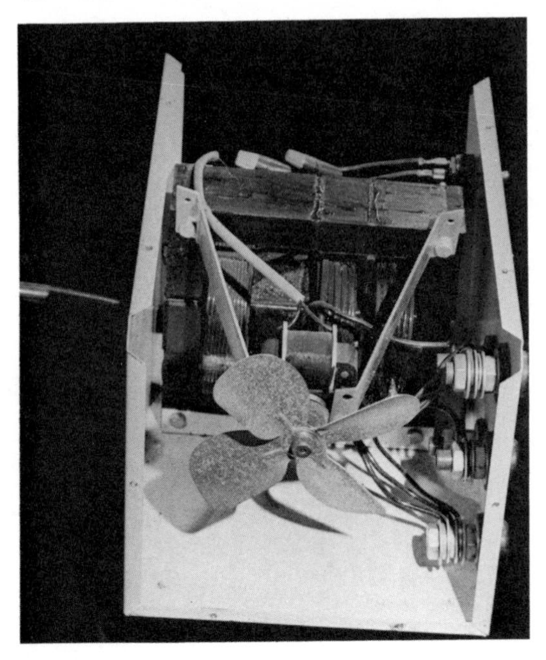

Internal parts of a bench-top welder are simple and rugged and require no maintenance. Iron plates with heavy wire wound around central leg make up the transformer. Wires lead out to selective amperage taps. Small fan is just for cooling.

How your AC welder works. Switch on your welder and you'll hear a hum. This is normal, and if you measured the frequency of the sound you'd find it was a 60-cycle tone. That's the alternating frequency of the line current into your house. It's reversing its direction 120 times each second and is called 60-cycle, or more recently 60-hertz, abbreviated 60-Hz. Without getting into more than the basic principles of electricity it's easy to see how, with the current flowing one direction, the wire-wound iron core becomes magnetized, with the north and south poles set up one way. Reverse the current and the magnetic field will collapse and reform with the poles disposed the opposite way. The magnetic field of the core reverses itself 120 times each second. The really interesting part is that this magnetic field (but not the input current from the house) is also passing through the second set of welder windings. This building and collapsing of the magnetic field "induces" a voltage and current in the output coil of the welder. The tricky thing is that you can change the relationship of voltage and amperage between the input and output coils of the welder by changing the number of turns in the coils.

If we wanted to produce extremely high voltage, as in an ignition coil, we'd use a few hundred turns of fairly heavy wire in the input (primary) winding and many thousands of turns of very fine wire in the output (secondary) windings. In a welder we want just the opposite: high amperage at modest voltage. For example, if we put in household power at 120 volts with the circuit limited by a fuse to 30 amps, we can take out 70 amps. But the maximum open circuit voltage will be only about 39 volts and the

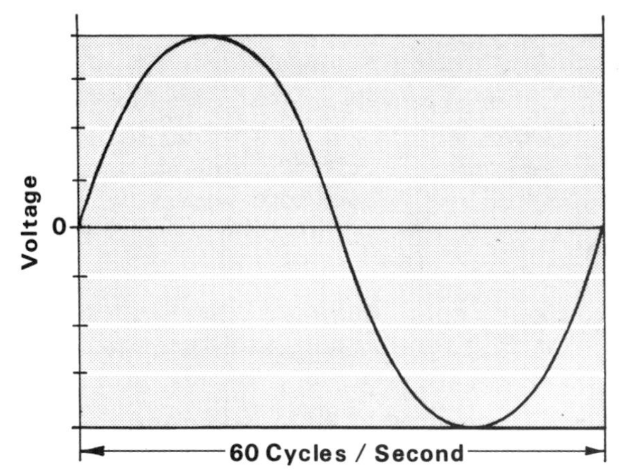

Secret of transformer action is alternating house current, which is visualized as rising and falling, and changing flow direction, along a curve called a sine wave. Action builds, collapses, and rebuilds magnetic field around transformer primary windings to induce voltage and current flow in secondary windings.

voltage while actually welding will be considerably less. We've made a trade-off—voltage for amperage.

What you need to know. If you've read all of the foregoing—and you couldn't be blamed for skipping it—there was a reason for my going to such great lengths to explain this voltage/amperage concept as it applies to arc welders. For one thing, it will help you in buying a welder. It will also help you install whatever you buy properly. And as you learn more

Primary windings, fed from house line, magnetize iron core of transformer. Primary and secondary windings are not electrically connected. Electrical input/output relationships reflect ratios between number of turns of wire in primary and number of turns in secondary.

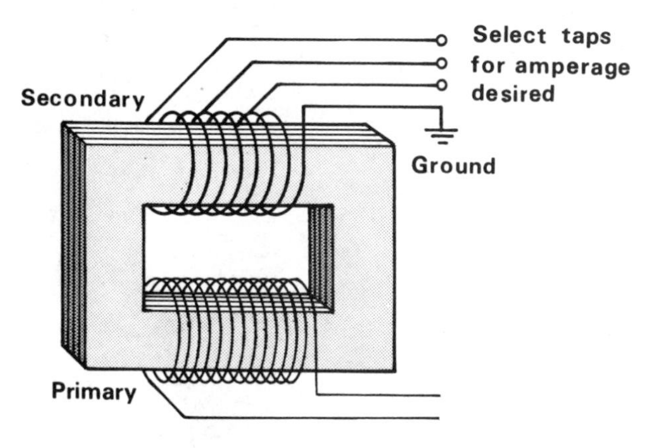

about welding and electrodes it will even influence the way you plan and design your welded joints. For those who didn't read, or didn't quite understand, how a transformer welder works, remember that the alternating input current also comes out as alternating welding current. This means you must use welding electrodes intended for AC, or AC/DC. Bring home some specialized DC rods from the plant where you work and you'll find you've wasted your time. All of the very light-duty and bench-top welders are strictly AC. More on DC later.

What capacity welder? Most important, when you buy a transformer welder you must give some thought to your household supply circuit. Before you do that you have to think a bit about just how heavy your welding workpieces are likely to be and what amperage you'll need. You can find charts telling you what thickness of metal can be handled by various rod sizes and amperages, but these tend to be based on ideal industrial results and don't take into account the ingenuity of home-shop weldors.

The lowest-amperage welder you'll find listed by a reputable merchandiser is 50 amps. Such welders have no adjustments. You plug them into a wall socket and get a maximum of 50 amps. What will 50 amps do? Well, you can expect to do a passing job on $1/16$" to $3/32$" steel, conduit and tubing,

Although not suitable for heavy welding, small, single-output, 50-amp transformer welder can be carried easily and plugged into standard wall outlet. With a little practice you can fusion-weld up to 3/32" steel, and carbon-arc braze and bend fairly heavy stock. Shown here being used as a welder.

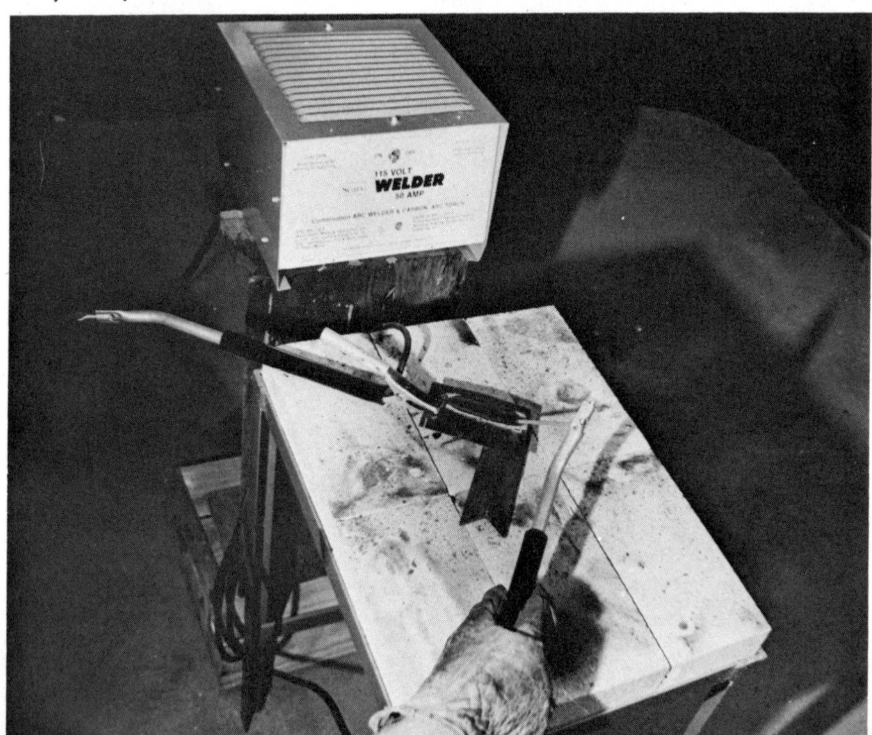

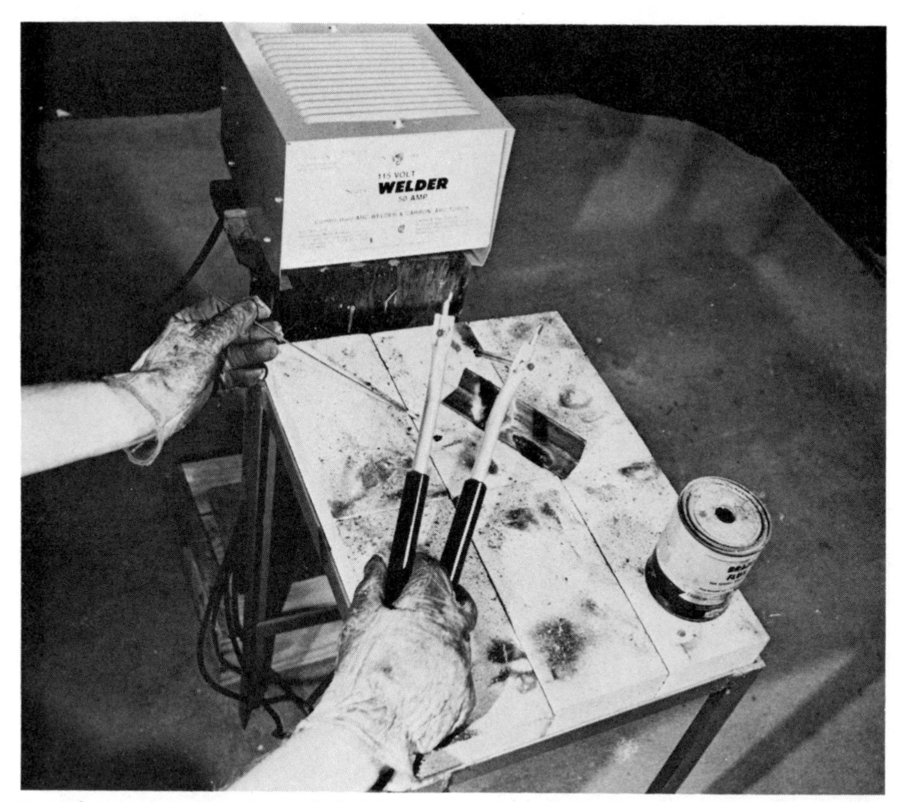

In carbon-arc mode, electrode holder and ground clamp for same 50-amp welder convert to a primitive carbon-arc torch. It puts out over 9000° of clean, inexpensive heat.

artifacts, and the like. The heaviest rod they'll handle well is 1/16″. Don't expect to weld 1/4″ plate, make swing sets or trailer hitches (which must be safe), or work on any equipment where a weld failure is serious.

Part of this is a matter of practice. I've had readers write complaining that 50-amp welders were a fraud; they couldn't weld with them. In reply I've sent them samples of welded metal made with the cheapest 50-amp welder on the market. But the facts are that you're working in a marginal area. It might be that their line voltage into the house was a bit low. Mine tends on the high side of 120 volts. It is much harder to strike and hold an arc with 50 amps than it is with 70 or 100 amps. Without some experience it's frustrating and with some electrodes almost impossible.

One thing you can do with 50 amps is use a carbon-arc torch for brazing and shaping metal. And, surprisingly, with a high-frequency converter you can do TIG welding very nicely. (TIG—that is, tungsten–inert gas—welding is an advanced form of arc welding used for aluminum, stainless steel, and other metals often difficult to weld conventionally.) Later I'll tell you about these tricks.

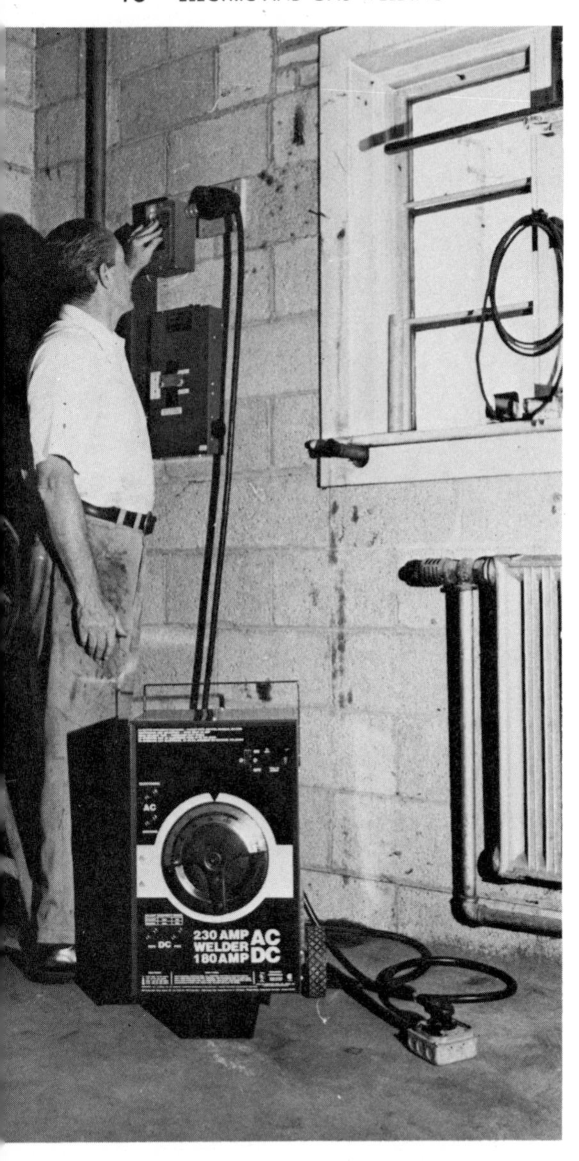

Upgrading to 200 or more amps from a 100-amp welder means that you'll probably have to install a separate 230-volt outlet with 60-amp fuses. Outlet shown here comes directly from distribution box and has a separate double-pole switch. Nothing else is on the circuit.

If I wanted to learn to do routine welding jobs, and if I didn't expect to get into large or technically demanding welding, my choice, without question, would be a bench-top welder that I could plug into a 30-amp house circuit and pull out 50 to 100 amps to weld. I'd want an adjustable amperage control or selective plug-in taps. Such a welder can be plugged in almost anywhere. It's light enough to carry about. And 100 amps will do most ordinary jobs with no problem.

230-volt welders. The next jump is a big one. When you get over 100 amps you'll find that you must plug into a 230-volt (sometimes 208-volt) circuit. You probably have such circuits in your house but they're dedicated to the stove, water heater, or maybe a clothes dryer. In some cases, if the laundry room is near the shop area, you can plug into the dryer circuit, but you'll almost certainly blow fuses whenever you try to use higher amperages. That's because dryers often have two 30-amp or 40-amp fuses, and if you look at the catalogue descriptions of 230-volt welders you'll see that they require 50-amp or 60-amp fuses. If you change your dryer fuses you are then running it essentially unfused, and that could start a fire.

A very important point is that the heavy welder needs a separate circuit! So do the 100-amp units! You don't want an air conditioner, well pump, or the like cutting in at the same time you're welding.

If you happen to have an unused 230-volt fuse block in your home distribution panel it's easy enough to bring out, or have an electrician bring out, a special plug for your welder. If your basic home wiring is low on muscle you may find that you need a major rewiring job to bring in heavier service connections and a new panel. That's why it's so important to check before you buy.

What about duty cycle? Duty cycle is another fine-print listing you'll find in the descriptive literature for electric welders. Some, for example, are 20%; others vary from 20% to 100% depending upon the amperage you're using. What the manufacturer is telling you is that you can use the welder only for limited times without allowing it to cool. A mail-order house describes the duty cycle for their welders as the ratio of welding time to cooling time per ten-minute cycle. I've never really tried to work that one out, primarily because I don't know any home-shop weldors who work on ten-minute cycles.

In reality, the average home user of light arc welding equipment normally finds himself using no more than a part of an electrode at a time. Then the workpiece has to be repositioned, or you have to reposition yourself, or you back off to contemplate what you've done and where you're going. I've used 20% duty cycle welders for years and can't recall ever exceeding the duty cycle.

DC WELDERS. In the early days of arc welding, practically all welders used direct current (DC). Direct current, of course, is like current from a battery. It doesn't have the 60-Hz alternation characteristics of house current. Also, DC will not work in a transformer. The extensive use of DC originally reflected the state of the art. The proper electrode coatings had not been developed for AC welding, and modern high-temperature insulating materials were not available for transformers. Since the only common source for direct current was batteries (impractical for welding),

DC had to be produced on the site. This meant that you had to have a DC generator driven by either a motor or an engine. Such equipment is still widely used for commercial welding, and you'll commonly see it in weldors' pickup trucks.

Commercial DC units, either engine-driven or motor-driven, tend to be expensive, very heavy, and impractical for home use. But recently small, very portable, single-cyclinder engine-driven welders using automotive-type alternators have become available. Typically, such a unit will have an 8-hp engine, weigh a bit less than 100 pounds, and provide about 135 amps, DC. They're ideal for welding away from power lines. Some will also provide power to run drills, saws, lights, and other equipment as emergency power generators. You can expect them to cost $500 or more.

AC vs. DC. You may wonder why I even mention DC welding. Is it better than AC? Can it do things AC can't? My answer—and this is strictly from the standpoint of the home-shop weldor—is a qualified "sometimes."

In describing AC welding I emphasized the reversal of current 120

Portable DC engine-driven welder can be carried anywhere; lets you weld away from power lines. Eight-hp unit shown delivers a maximum of about 135 amps, enough for welding fairly heavy steel.

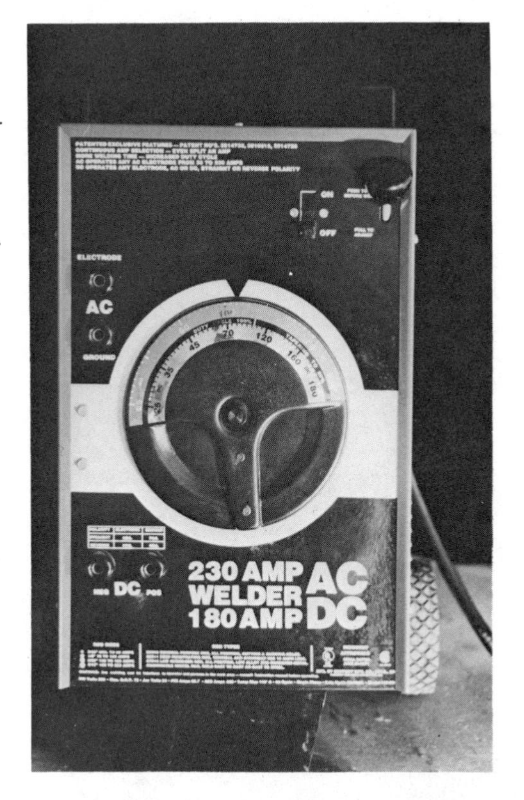

By plugging into AC, top, or DC, lower, taps, you can select the performance characteristics you need from this basic AC transformer welder. Built-in rectifier converts AC to DC, but maximum output is lower. DC option is well worth slight extra cost when you buy.

times each second. The same reversal action occurs in the arc itself. The electrons in the arc gap are flowing first one way and then the other. In some ways that's desirable. The alternations are said to break up and carry scale from the weld surface. More important, the flow of electrons across the arc gap establishes a magnetic field, just as in any other conductor. With AC this field reverses polarity constantly. With DC the polarity remains fixed. Many times, especially when welding in a corner or a pocket, the DC welding arc will tend to be attracted and diverted from where you want it by magnetic action. This effect is not a problem with AC.

This single-direction current flow can also be put to advantage with DC. You'll hear professional weldors speak of "straight polarity" or "reverse polarity." According to electrical theory, current flow is the movement of electrons. Electrons flow from negative to positive. Thus if you connect the electrode holder cable of a DC welder to the negative outlet the electrons will flow from the electrode to the workpiece. The latter, of course, is connected to the positive outlet to complete the circuit. This is called straight-polarity welding. It is claimed that with straight polarity one-third of the total heat is released at the electrode and two-thirds is released at the workpiece.

With reversed polarity, the positive outlet of the welder is connected to the electrode holder and the negative outlet to the workpiece. Now, one-third of the heat is released at the workpiece and two-thirds at the electrode. This, it is claimed, results in superheating the electrode and shield material so that the arc has a very high velocity and the metal is projected across the arc at high speed and with great impact into the molten puddle. Theory or not, with DC you can vary the penetration into the work metal. Essentially, this makes vertical and overhead welding easier.

In professional welding this can be very important. It may be less important in the home shop, where vertical and overhead welds are less common. On the other hand, there are some very handy tricks best done with DC, among them rivet welding and thin-metal welding, that make DC extremely attractive. Fortunately, the latest welders in the larger sizes are offered with an internal rectifier and are called AC/DC machines.

CABLES AND ELECTRODE HOLDERS. Most home welders will come with a pair of cables as standard equipment. One cable will be fitted with a heavy metal spring-operated clamp called a ground clamp. The other cable will have an insulated clamp with jaws to accept the welding electrodes. Although there are many variations in cables and clamps from

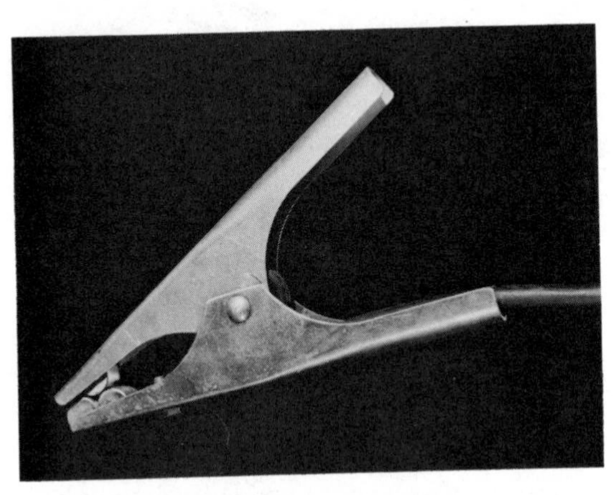

Ground clamp is a vital part of arc welding circuit and must make firm electrical contact for proper performance. Note the heavy copper jaws on this clamp. Spring action is quite stiff, but even so you should take care to get a really secure grip.

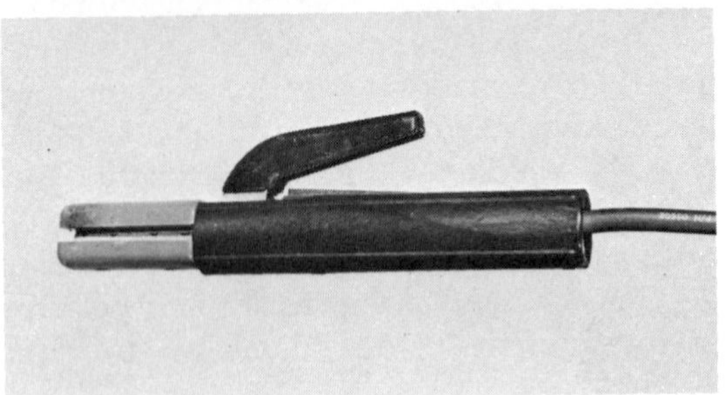

Heavily insulated electrode holder must lock electrode snugly but release instantly. Again, heavy copper jaws are needed. When buying, check to see that cables are labeled "welding cable."

You don't want cable terminals that accidentally pull loose while you're welding. Loose connections can damage welder. When buying, test for secure gripping or locking action of terminals. Before welding, push them in tightly.

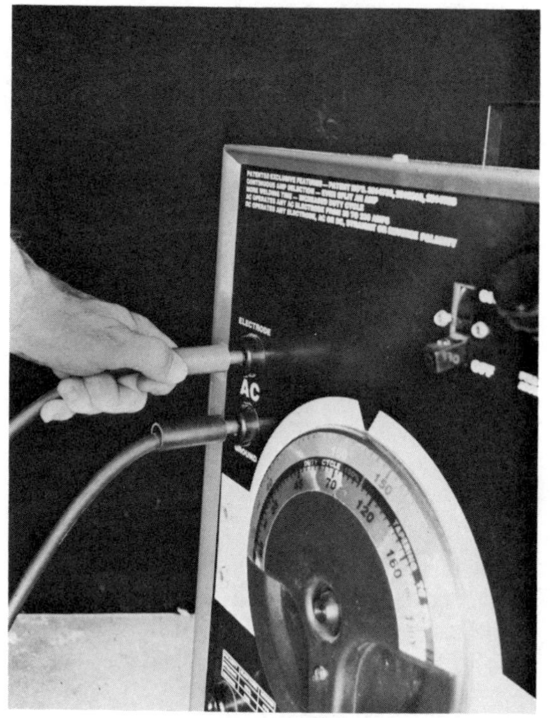

different manufacturers, their overall quality is a fair guide to the quality of the welder. Naturally, you can't expect the finest professional cables and clamps with low-cost home-shop welders. After all, you can always replace them with something better if you wish. As a general guide, however, here are things to look for:

- The cables should be of copper construction and very flexible. Some cables are made of aluminum, but its current-conducting ability is lower and it tends to stiffen as you use it.
- Look for markings on the insulation that designate the cable as welding cable.
- The ground clamp should have husky copper pads to make good electrical contact with the workpiece.
- Look for a good copper gripping surface on both jaws of the electrode holder. Try clamping a rod in the jaws and wiggling it. The grip should be very secure but capable of instant release.
- Check for a twist lock or very secure taper fit where the cables fit into the terminal taps. Loose fittings here result in cables pulling loose while welding, which can cause burning of the tap sockets and can damage the welder internally.

3 Other Arc Welding Equipment

As with nearly any technology, you can expand your working range in arc welding with a variety of add-on equipment. First, however, let's examine the basic protective gear you need for your own safety and then look at the items that are "nice" to have and very often fun to use.

EYE PROTECTION. If you buy a new welder, even an inexpensive one, it's probable that you'll get some sort of a helmet with a dark viewing window as part of the package. If you don't get one, don't even plug in your new welder until you do. *Above all, don't try to get by with gas welding goggles or other substitutes. You'll damage your eyes and get severe ultraviolet burns on your face.*

Almost any helmet will do for occasional light welding below 70 amps, but often the lens furnished with light welders is too transparent for safe use. In general, you'll find a #10 lens suitable for rod from 1/16" to 5/32" diameter. Never use a lighter one. For heavier work a #11 or #12 is better. For carbon-arc work it is a good idea to go to a #14. If your eyes itch and feel gritty after welding it's a sure sign your eye protection is not all it should be. Your eyes have been sunburned, so to speak, by the ultraviolet from the arc.

Anything which interferes with your visibility or comfort while welding makes it that much harder to learn or do a good job. A shoddy helmet can be very annoying. Suppose you're trying to put the workpieces in approximate position and the helmet flops down over your face; or suppose you flip back the face piece and a sticky pivot at the headband results in your flipping the entire helmet off your head. Helmets that don't adjust properly and stay put have a habit of slipping down and cutting off your vision right at a critical moment while welding. Worse, when you get some of them adjusted so you can see properly you find you've got a gap just below

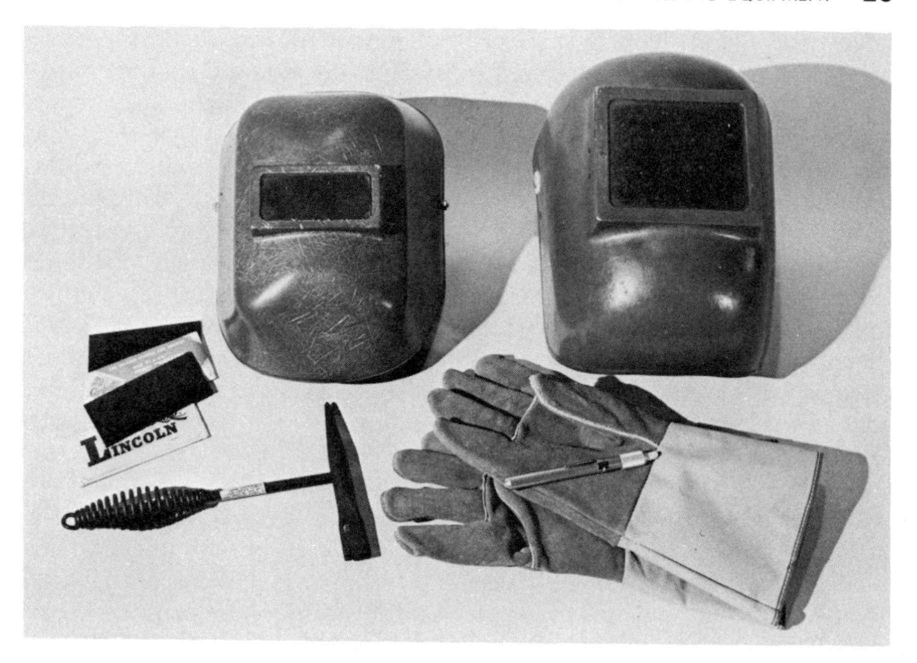

Protective helmet and flame-resistant gloves are important to your safety. Standard, narrow-lens helmet, left, is hard to use with eyeglasses. Wider lens, right, is better. Dark lenses come in different grades, left; choose to suit your work. The chipping hammer and soapstone marker are handy accessories.

the chin which allows hot sparks and bits of metal to come up under the helmet and chase around inside. A better-grade helmet doesn't cost much, and I recommend buying a good one when you buy your welder if there's an obvious cheapness about the one that comes with it.

If you wear glasses. Those who wear bifocal or trifocal glasses will soon find that it's extremely hard to line up the lower or close-in portion of the eyeglass lens with the helmet window after you strike an arc. There are two solutions to this common problem. One is to buy a helmet with a big, wide viewing window rather than the little narrow window that's more or less standard. The other is to provide either a pair of cheap "reading" glasses or add a magnifier lens to the viewing window. Such lenses are available from welding supply houses. They work well with bifocals, since you can look out through the portion of your glasses above the close-up section and the magnifier does the rest. With trifocals there's another special section, however, right where you want to look. Reading glasses are the only answer.

AVOIDING BURNS. It might be considered macho to arc-weld barehanded with an open shirt and sandals, but there's nothing macho about skin cancer and deep burns from hot metal. All arc welders produce intense ultraviolet rays, and in a short time any exposed skin acquires a brilliant sunburn. This is more than just a nuisance. Such exposure will sooner or later cause skin cancer on most persons. A good pair of treated, flame-resistant gloves made for welding, a wool shirt (try military surplus) that buttons right up to the collar, a helmet that curves in well around the sides and under the chin, and heavy shoes are cheap protection. A leather apron is not a necessity but it certainly helps if you sit down to a bench to weld. Beware of slacks or shirts, as well as gloves and caps, which are made of modern synthetic fabrics or imitation leather. Most of these will burn with a creeping characteristic if ignited by a spark. Some will burn furiously. All are dangerous. Even so-called work pants today are often a synthetic blend and a few minutes of welding can easily produce "measles of the thighs."

ELECTRICAL PROTECTION. I always try to wear heavy work shoes with rubber soles when electric welding. This is an important protection against electrical shock. As discussed earlier, the typical open-circuit (when you're not welding) voltage on a transformer welder ranges from 40 to 80 volts. Normally you can insert an electrode into the holder and feel nothing at all, especially if you're wearing heavy gloves. But sometimes, if you have thin-soled shoes and the ground or shop floor is damp, you'll get a tingle. Common sense points to wearing shoes with good insulating qualities.

There are always possibilities for shock from any electrical device, including household appliances and shop machinery. Your arc welder is perfectly capable of developing a short from something as simple as a chafed input cord, broken-down internal insulation, or even a scrap of metal which falls inside through a vent opening. All welders from reputable manufacturers have three-wire input leads and plugs and are grounded internally back through the third wire. But some users will persist in using them with cheater plugs, or the house circuit may not be grounded, as is common in older homes. If your welder comes with an external ground wire and the instructions call for attaching this to a water pipe or other known ground, do not overlook doing so. The object is to prevent shock from touching the welder housing if a short should develop. If your welder has no provision for such a ground, it costs nothing to add one by clamping a fairly heavy flexible wire under any convenient external screwhead or bolt and running the other end to a grounded pipe or conduit. A handy device to secure the wire to the pipe is a small hose clamp such as used for car heater hoses.

CARBON-ARC—BEST HEAT IN THE HOUSE. Not many of us can recall the old "arc" streetlights, but carbon arcs are still used in some types of searchlights. When two pieces of carbon are connected to opposite sides

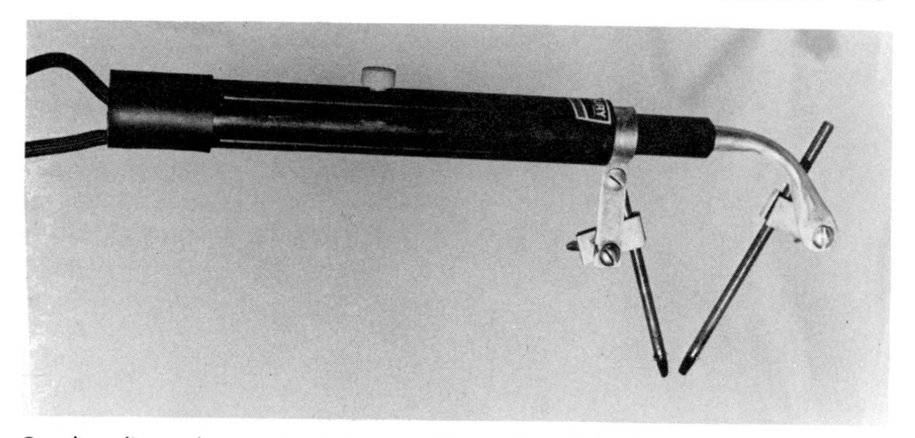

Good-quality carbon-arc torch has a solid handle, adjustments for angle of carbons, and a smooth-working thumb control for easy arc starting and adjustment. Carbons are coated with copper for good electrical conduction.

of an electrical circuit and touched together, a brilliant blue-white light results. This light is very high in ultraviolet and absolutely requires eye and skin protection—something the old-timers overlooked, to their regret.

From the shop standpoint, the important thing is that a carbon arc is probably the most intense heat you can produce in your shop. Reportedly it approaches 10,000° F. Surprisingly, carbon arc is also your cheapest source of heat. Suppose you want to bend some heavy rod, pipe, or scrap iron, or put a neat twist in some ornamental iron. You can waste a lot of time and gas with a propane or MAPP torch, or spend less time but more money with oxyacetylene, or you can plug a carbon-arc torch into any transformer welder, even a small one, touch the carbons together, and in seconds bring the metal to a bright bending heat. I point out that even little welders work fairly well, since you'll seldom need over 65 amps for most jobs.

Or suppose that you want to braze-weld two pieces together, or even weld aluminum. The carbon-arc torch is quick; the carbons cost only pennies, and the electrical cost is trivial for most jobs. Later, we'll take a look at the actual techniques, but for right now let's look at what a carbon-arc torch consists of.

Carbon-arc torches. In its simplest form the cheap little electrode holder that comes with some small welders converts into a carbon-arc torch. If you're good at eating with chopsticks you'll probably get along fine with this outfit, since it resembles a big wishbone and you control the arc gap by squeezing the sides together with one hand. It works, but a much better carbon holder is available for about $20 and has the weight and feel of a good tool. Most important, such torches have a sliding thumb button for

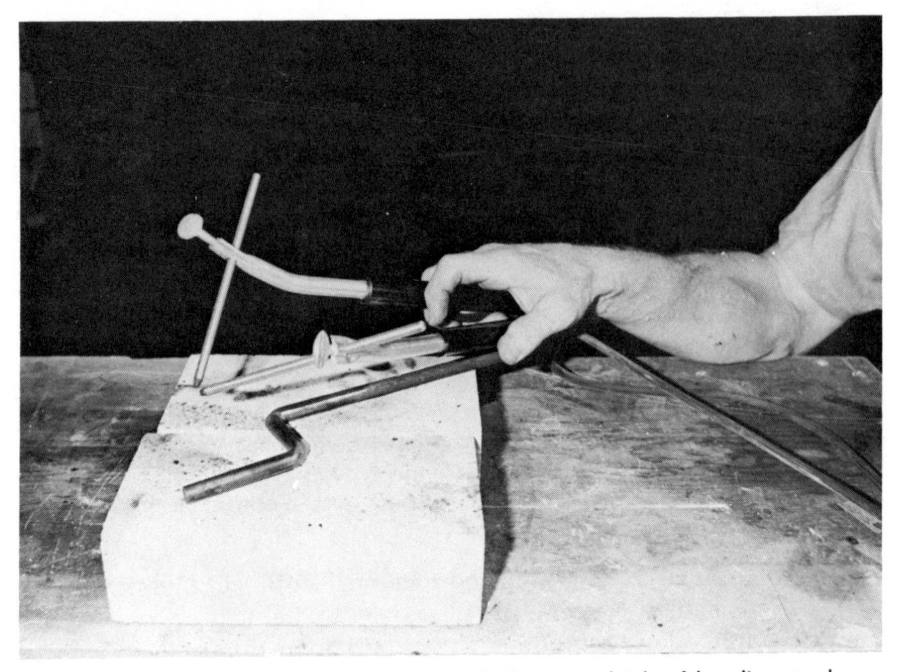

Even little "chopstick" carbon-arc torches will do a good job of bending steel such as this ½" rod. Lacking precision arc adjustment, they are harder to use for brazing than better torch with locking thumb adjustment.

touching the carbons together and then adjusting the arc. Many owners will keep such a torch plugged into a bench-top welder and ready for action at any time.

Carbons for AC and DC are formulated and constructed just a bit differently. Be sure to buy the type you need. They come in ¼", 5/16", and ⅜" diameters; 5/16" is probably the handiest.

HIGH-FREQUENCY CONVERTERS. When you buy your first AC transformer welder and start learning to weld, the more exotic forms of welding will probably seem totally remote from your interest. Later, you'll start browsing through catalogues and find that there are some extremely interesting things you can use to upgrade even a simple 50-amp "sputter-box." The basis for this is a high-frequency converter that will let you weld aluminum, stainless steel, and other hard-to-weld materials. A converter raises the alternating frequency of your arc discharge to a level where the arc literally leaps to the workpiece without your having to strike an arc. You don't have to touch the electrode to the work metal.

Be assured that your first efforts at arc welding will be frustrated by the inability to strike an arc and hold it. We'll talk about that later, but basically, you strike an arc by touching the workpiece briskly with the electrode tip. You can use either a sweeping or a tapping motion. The problem arises when you have two pieces of metal propped or lightly clamped exactly the way you want them. Striking blindly, as you must with a welding helmet, the chances are very good that you'll knock your pieces out of alignment, or the rod will stick and you'll ruin the entire set-up, breaking it loose.

With a high-frequency converter, often called an arc stabilizer, you can sneak the electrode tip into the work area, let the arc jump to the workpiece, and quickly bring the arc to the desired location without touching the work. This helps greatly with light fit-ups.

TIG WELDING. Easy arc starting is not the only thing you can do with high frequency. Another accessory called a TIG (tungsten–inert gas) torch will let you do some fairly exotic welding, such as repairing aluminum castings, joining aluminum and stainless-steel assemblies, and even welding Monel, titanium, magnesium, and other unusual metals. Since no flux-coated rod is used and the weld is shielded by inert gas, there is a minimum of contamination and an extremely pure metal is deposited in the joint. The package consists of a set of gages and a gas regulator, a hose, and a torch with a small tungsten metal electrode inside. The nose of the torch

TIG welding, using a high-frequency converter, left, is possible even with small 50-amp basic welder, right. Argon tank and regulator feeds inert gas to torch clamped in electrode holder. With TIG, aluminum welding suddenly becomes easy.

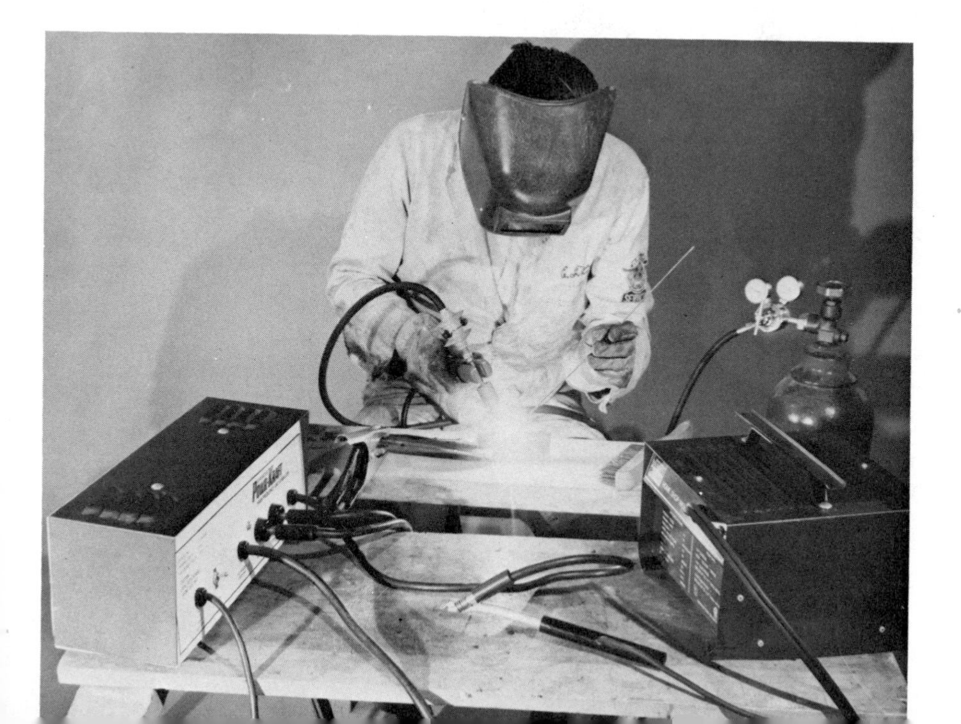

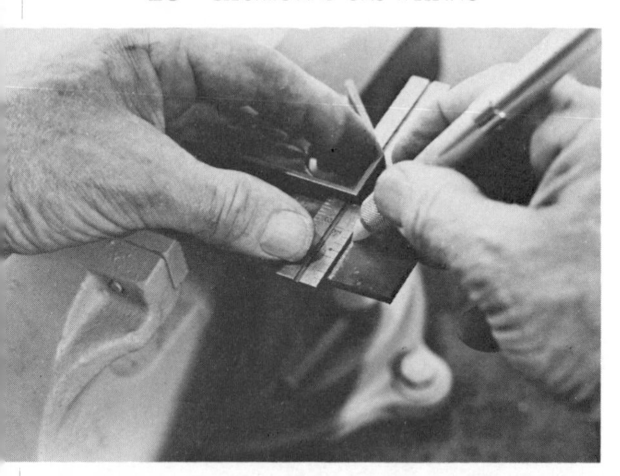

Soapstone marker is weldor's tool for laying out cuts and marking reference points on dark metal. Start working accurately from the beginning. Welded metal is hard to change if you go wrong.

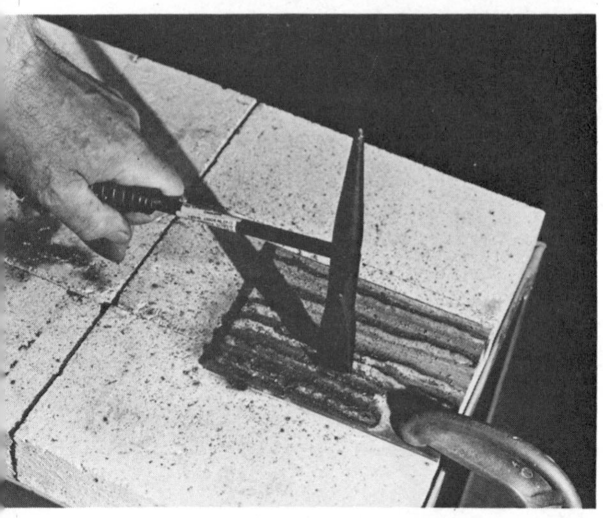

Chipping hammer is best tool for rough removal of slag formed from electrode coating. Never use it without eye protection.

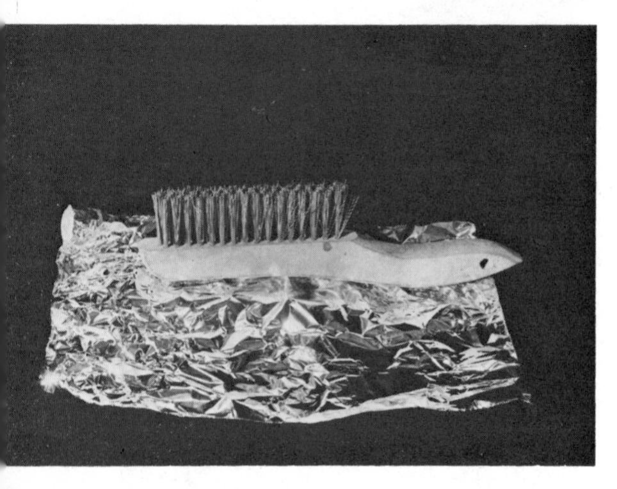

Ordinary wire brush is needed to clean up welds. A stainless-steel brush is required to clean aluminum before welding; keep it meticulously clean and wrap it in aluminum foil and bag for storage.

is a ceramic cone. You'll also have to rent a tank of argon gas. Argon is inert and noncombustible.

In use, you hold the torch in the standard electrode holder and adjust the regulator to flow a small amount of argon gas out of the ceramic cup around the needlelike tungsten electrode. Since the gas is inert and heavier than air, it envelops and shields the work metal from the air to prevent oxidation. The arc leaps to the metal and forms a puddle, and you feed in filler rod as needed with your left hand. The tungsten electrode does not melt or enter into the operation except as a conductor of the high-frequency current.

An aluminum weld made by the TIG process can be a thing of beauty, and we'll talk more about it later. The important thing is that high-frequency equipment can be added to any basic AC welder at any time.

SMALL WELDING TOOLS. Working with the usual black or dark metal you'll use for most projects requires some sort of layout marker other than a carpenter's pencil. When following a line viewed through a dark helmet lens it helps greatly if the line is white. Traditionally, for practical reasons, weldors and metal workers use a soapstone marker. The soapstone is inexpensive and even comes as an insert for a pencil-like holder. Since you'll be trying to follow lines as you practice, I recommend getting in the habit of using such a marker.

As mentioned, electrodes are heavily coated with a flux material which melts and leaves a protective slag over the weld. You must chip this slag away after welding and especially before starting a new weld over an existing weld, as happens when you have to stop and put a new rod in the holder. Almost anything will whack the slag away after a fashion, but a standard weldor's chipping hammer works better.

A stiff wire brush is also needed for clean-up before and after welding. Such clean-up is very important if you're going to paint your work, since the residual slag will eventually loosen and ruin the paint and appearance of the job. Although a wire brush is adequate for most work, you may have objects which will be on display as furniture, such as the table shown on page 7 in this book, or as art objects. For such projects it's well worthwhile to take the finished project to a commercial sandblast shop for cleaning to a bright scale-free and slag-free surface for priming and painting. Another option, especially handy for automotive and aircraft work, is a small, portable sandblast gun which operates from a heavy-duty paint spray compressor.

If you plan to weld aluminum, just any old wire brush won't do. The usual brush introduces contamination when you brighten the weld area before welding, and for this reason you need a stainless-steel brush. This brush must never be used for anything but aluminum, and when not in use it should be wrapped in aluminum foil and slipped into a plastic bag to keep it free of shop dust.

4 Arc Welding Electrodes

A complete discussion of all the arc welding electrode types on the market would require at least a full book of this size, leave you completely confused, and be out of date before the book was printed. The reason for this proliferation of apparently "alike" products is not only competition among electrode makers but also a real, ongoing need for better, higher-production welding at the industrial level. Read a few of the electrode descriptions and you'll soon discover that many of them are designed to work only on DC, reversed polarity, or that they have special penetration characteristics, are needed for high-alloy steels or to minimize slag deposits in deep crevice welds, or that they have other special features for high-production industrial welding. From the standpoint of the home-shop weldor it's seldom, if ever, that you'll find a real need for exotic electrodes.

First of all, an electrode is a metal rod. Usually it will have a composition quite similar to that of the workpiece. For example, to weld mild steel you use a mild steel rod. The coating on the rod is extremely important, since it not only shields the molten metal from the oxygen in the air but also has a profound effect on the arc characteristics, such as arc starting, stability, and length. If you want to experiment, try to strike an arc on a light-duty AC welder with a rod made exclusively for reverse-polarity DC.

Fortunately, most electrodes, in addition to carrying their maker's product names, are also numbered according to an AWS (American Welding Society) standardized code. As an example, the rod you'll almost certainly use much of the time will be an E-6013. This is a conventional mild steel rod. The E is for electric welding. In the four-digit number, the first two digits, 60, indicate a tensile strength of 60,000 pounds per square inch. The third digit, 1, signifies that it may be used in any position. The fourth digit, 3, means that it's suitable for either AC or DC and the latter may be either polarity. You'll usually find such a number on the holder end of each rod. Sometimes you'll also find a color-code dot. The color code is not well

Most standard electrodes have basic AWS identification numbers on the shank. Top three, 7014, 6013, and 6011, are ones most commonly used in home shop. Lower, 308-16, is special for stainless steel.

established, however, except that an E-6013 electrode will carry a brown dot.

If that were all there is to it, choosing the right electrode would be fairly simple. Unfortunately, electrode makers try so hard to make exactly what their big industrial customers want that they complicate things for the home-shop weldor. The following is just one example of rods having the same AWS number showing very different performances.

Earlier we compared welding to freezing water, and I mentioned that some electrodes are classified according to their "freezing" characteristics. This becomes important, for example, in vertical welding when we're

Electrodes are more complex than they appear. Basically, central rod conducts current to arc gap, and coating melts and vaporizes to form a shield around molten filler. Melted coating leaves slag to protect metal as puddle moves along the joint.

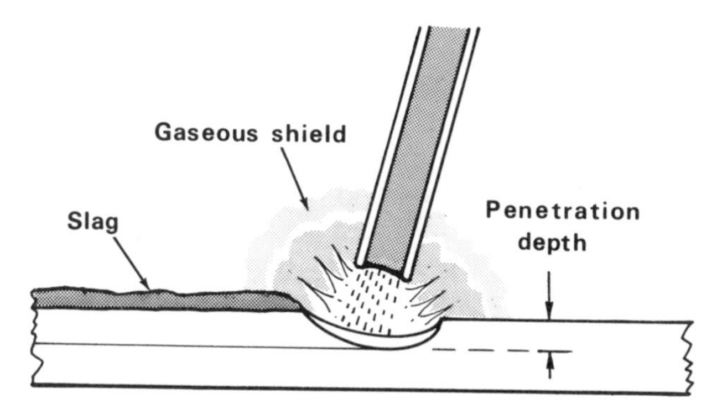

trying to keep molten metal from flowing downhill. Ideally, logic tells us that we would want a so-called "fast-freeze" electrode which solidifies quickly and doesn't run downhill easily. But selecting such a rod is complicated by products which basically fill the specifications for standard designations and are so labeled but which also are made in variations which perform differently.

Thus, referring to a Lincoln electrode listing we find that the Fleetweld 37 rod is "especially suited for downhill and vertical-down sheet-metal welding using a drag technique." The AWS number for this rod is 6013, and that's marked on the rod. Fine, that's our old standby, E-6013, that I mentioned above. But wait, it happens that Lincoln's Fleetweld 57 is also a 6013 rod and is so marked, but when it comes to downhand welding it's recommended for a maximum of only 30°. The first type 37 rod would work fine for vertical welding but the type 57 would be troublesome. It should also be noted that the type 37 can be used with AC or DC of either polarity. The type 57 is marked for AC or only straight-polarity DC.

Continuing our examination of available rods for vertical welding—and these are only examples—we find that for AC/DC a 6011 rod is a general-purpose, fast-freeze type described as "particularly good for vertical and overhead." We also find that the basic 6011 rod is offered by Lincoln in three variations, all of which are AC/DC, but one of which, Fleetweld 35, is for reverse polarity only on DC.

My purpose in leading you into this maze of electrode listings is not to confuse you but to point out that the above examples are only a quick look at one manufacturer's listings and that there are dozens of electrode products on the market. About the only recourse the home-shop weldor has is to try to confine most of his work to more or less straight and level positioning, and, secondly, to get several product listings from his supply shop and study them carefully. This will at least let you select and try a few electrodes that sound right for your purposes.

Obviously, an experienced weldor, or the welding supply salesman, can be asked for advice; but, be warned—they think differently than you do. Read the descriptions of various electrodes and you'll often discover that when they talk of vertical or off-position welding they're thinking in terms of ¼" to ¾" plate or pipeline welding. Most home welding tends to fall into what the pros call sheet metal. You may not think of ⅛" or 3/16" steel as sheet metal, but they do. They'll make suggestions, but quite often if you read the literature you'll discover that although some of their favorite rods are for thinner stuff, most are intended for heavy industrial plate.

To further muddy the waters, when you go out to buy rod you'll run into the fact that although your light-duty welder requires, or at least works best with, 1/16" electrodes, some manufacturers do not make them smaller than 3/32" diameter. This is another reason I suggest E-6013 for small welders and general-purpose use. You can get it in 1/16", 5/64", 3/32", ⅛", 5/32", 3/16", 7/32", and ¼" diameters. Other rods, particularly E-6011, are also

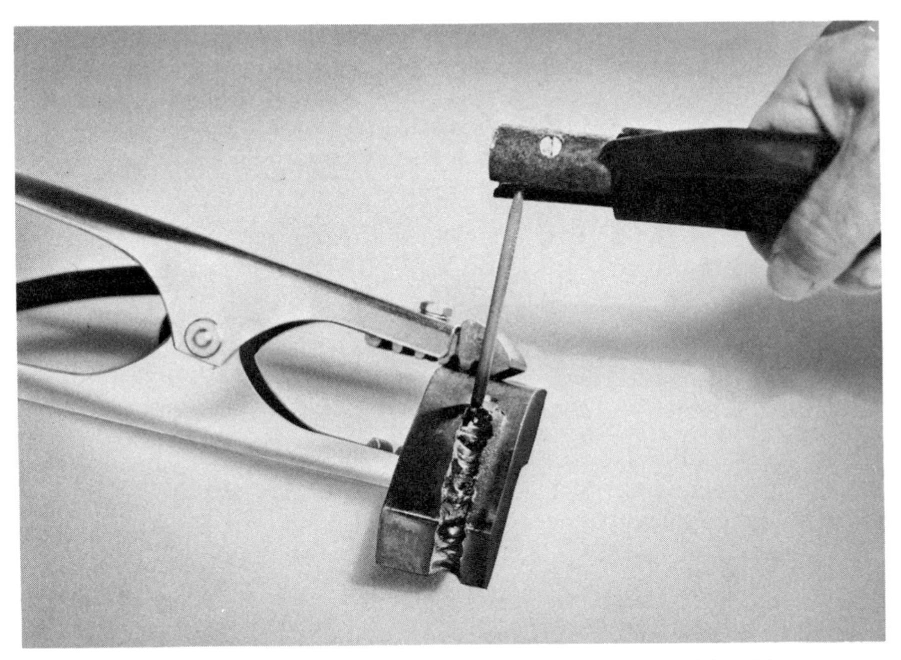

Cutting rod acts almost like a powerful chisel to gouge out an old weld or open a crack for welding. It leaves a clean work surface.

good, and the 5/64" minimum diameter can be handled by most small welders. With E-6011 and the single-amperage 50-amp models you may have trouble striking an arc.

BEST RODS FOR THE HOME SHOP. To cut through the confusion and give the beginner something to work with, here are some electrodes with basic characteristics which will handle nearly any ordinary home-shop job. The numbers given are the AWS designations I've mentioned. To avoid getting into messy problems with the variations of different manufacturers I suggest buying your electrodes at a mail-order or chain hardware store. They can't be bothered with big inventories to cater to niceties and sell pretty much standardized electrodes.

E-6011 Mild Steel AC/DC (reverse polarity): Flux coating made especially for dirty, rusted, or galvanized metal. Deep-penetrating, fast-freeze, good arc stability, good general-use electrode.

E-6013 Mild Steel AC/DC: Best all-around, all-position rod, moderate penetration. Good for lighter-gage steel. Easy arc starting and handles wide amperage ranges well. Puddle solidifies quickly.

E-7014 Mild Steel AC/DC: Easy starting rod for the beginner. Has powdered iron in the coating. Works best with clean metal and good fit-up. Rod may be used in contact with workpiece.

Machineable Nickel AC/DC: For cast-iron repairs, joining cast iron to other metals. Weld easily machined or dressed. Work area should be cleaned by using cutting rod first. Peen weld during cooling.

Cutting Rod AC/DC: For removing old welds, cutting, piercing. Use like a lance or gouging tool at 15° angle. Coating helps keep rod cooler and leaves a clean surface for later welding.

E-7018 AC/DC Mild Steel Low Hydrogen: All-position rod especially suited for high-carbon steels, spring steels, and different metals in combination. Sensitive to moisture in the air. High tensile strength.

A5.3-69 Aluminum DC Reverse Polarity or with Carbon-arc Torch: Heavy flux coating protects against oxidation. May require preheat of work. Requires warm water and wire brush or acid rinse to remove flux. Very sensitive to moisture; not for off-position welding.

Hardfacing Rod: Many trade-name products for applying super-hard coating to plowshares, worn components, earth-engaging tools. May be AC or DC, but maker's recommendations should be followed.

5 Learning to Arc-Weld

Running an arc weld is like diving off a high board. Once the arc is struck and the metal is flowing you are more or less committed to proceed. Your helmet shuts out everything around you. If a spark starts a fire you probably won't know it unless you smell smoke. If a child wanders on the scene and stares at the arc you won't know it until eye damage has occurred. If you just get started and your set-up falls apart or distorts it will be difficult to correct. In short, your entire attention is focused on the glowing puddle of molten metal.

Check your power supply. The above points and many others make it important to put your welding house in order before you start. I'm going to presume that you've just purchased a bench-top 50–100-amp welder intended for use on 115–120 volts. Such rigs are designed for operation on a 30-amp fused household circuit. But they are not designed for operation on a sharing basis with a refrigerator, air conditioner, dishwasher, freezer, well pump, or other household equipment. They need their own circuit. Thus, you'll do well to try unscrewing some fuses or flipping off some circuit breakers to see what's on whatever circuit you plan to use. It may be that you have no choice but to turn off the air conditioner or wait until the dishes are washed before starting to weld.

As an example, I have three 200-watt lights above my welding area. Once I've made my set-up I turn 'em off. Another 60-watt bulb on a different circuit provides enough light, and since the big lights are on the welder circuit I'm pushing things to an extreme if I leave them on. An incidental, but practical, point here is that overhead lights tend to come in from the back of your helmet and reflect annoyingly in the lens.

Fire safety. The next step is to survey your work area for at least 10 feet around the scene of action for anything inflammable. Oil-soaked or paint-

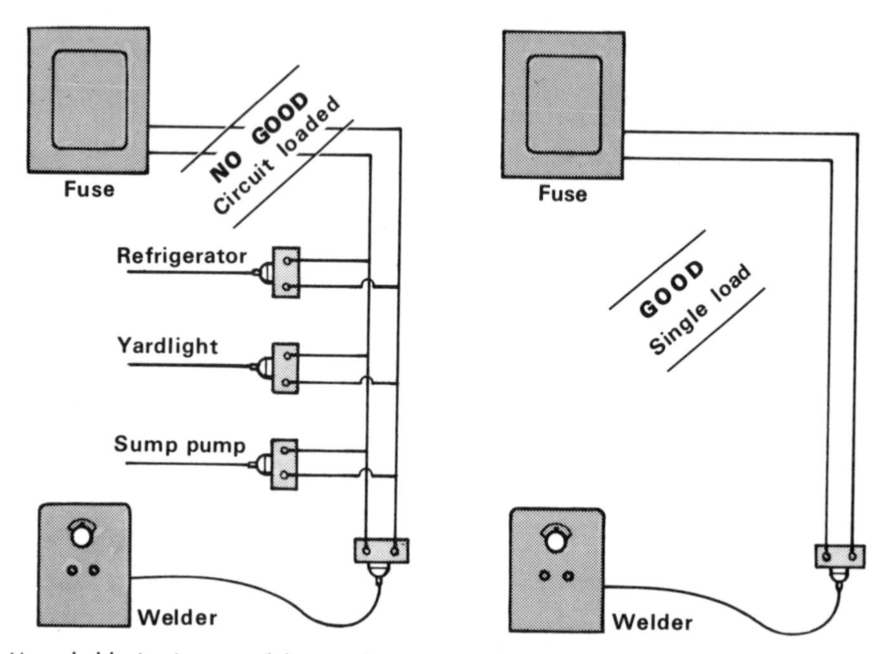

Household circuits can deliver only so much power without overheating dangerously. This is limited by a fuse or circuit breaker. Even small welders draw heavy power, and plugging into an already loaded circuit is not good practice. Locate a relatively load-free circuit, if you can, or run one in to supply your welder.

soaked rags; open cans of solvent, lacquer thinner, stains, or other liquids; wood shavings; and tools and mechanisms which won't tolerate hot metal spatter should all be removed. An example would be a set of micrometers or a precision vernier scale which could be ruined by an adhering globule of metal. Don't overlook your shop wastebasket. One glowing fragment of metal can sneak in and hours later cause a flareup in the usual junk, shavings, and old papers and wiping rags common to such receptacles.

Give some thought to the floor. Magazines like to publish lovely shop pictures showing tile, vinyl, or painted floors with Momma and Poppa sharing amazingly clean hobbies. The other extreme is the cobwebby old craftsman's shop in a barn with a wooden floor and burnable antiques hanging all around. Unfortunately, welding doesn't fit either picture. Be assured that molten weld spatter will shower the floor. You'll shake out red-hot electrode stubs on the floor, and there will be sharp-edged cuttings and ground-off metal underfoot. The fancy floors will soon have more burns than surround a barroom spittoon, and the wooden floors can harbor sparks that will send the whole shop up in smoke.

Your work surface should be totally fireproof, preferably firebrick or metal. Trying to weld on a wooden bench top will sooner or later cause a fire. Trying to substitute concrete block, ordinary brick, or tiles for firebrick is downright dangerous. They all explode with dangerous force under the intense heat of an arc. Firebrick is inexpensive, and a plan at the end of this book shows you how to make a simple angle-iron bench with a firebrick work top.

The final thing you'll need to get together is plenty of practice metal. Any scrap angle iron, sheet steel, bars, or whatever you can pick up will serve, but for your first efforts try to get some 3/32″ or 1/8″ sheet steel. If you're starting with a 50-amp welder, 1/16″ is OK. Pieces from 5″ to 12″ long are about right, and they should be reasonably clean of rust or paint. Do not use galvanized or cadmium-plated material.

YOUR FIRST ARC WELDS. When you've gone through all of the motions above and, I hope, have read your welder's instruction manual, you're ready for the big plunge. Clamp your practice piece to the table firmly so it can't move, as shown in the photo. Put an E-6013 rod, 1/16″ or 3/32″ diameter, in the holder, clamp the ground clamp snugly on the work, and flip the welder switch on. Focus your attention on the exact spot you want to strike the arc and flip down your helmet. As a last admonishment, if you have a cigarette or a pipe in your mouth, remove it. Flipping a heavy

Firebrick work surface on home-built welding bench resists arc and flame and will not explode as ordinary bricks or concrete will when heated. Set-up here is for practicing welding beads.

helmet down with your mouth so occupied isn't as funny as it sounds.
From here on a number of things will certainly happen:

- You won't be able to see a thing.
- You'll stab with the rod and not even hit the workpiece.
- You'll hit the workpiece but not where you intended.
- The rod will stick to the work and glow red.
- You'll get a sputter and a flash but won't be able to hold an arc.
- You'll decide that arc welding is impossible.

Anyone who says that these things didn't happen to him when he first tried arc welding is charitably given credit for a bad memory. Every arc weldor has gone through these experiences. Unfortunately, after a few hours' practice the memories are buried and they can't understand why the other guy has trouble. No matter; throw away the rod if it stuck and turned red. Next time you'll know that a quick flick of the wrist will usually snap it free. Or you can release the holder clamp and break it free when it cools. Sticking is routine for the beginner and even for the advanced amateur. You'll soon learn to sense it and break it automatically as a reflex action.

You'll also learn to hit where you aim on the workpiece fairly well. After all, you learned to eat with a fork and you can't see your mouth either.

Striking the arc. My own suggestion for striking the arc is to flick the tip of the rod across the work briskly, much like striking a kitchen match. Many pros simply tap the rod on the metal and draw an arc. That's fine for them, and for you, too, if it seems easier, but my experience is that dragging the tip on the metal works better when working with low amperage.

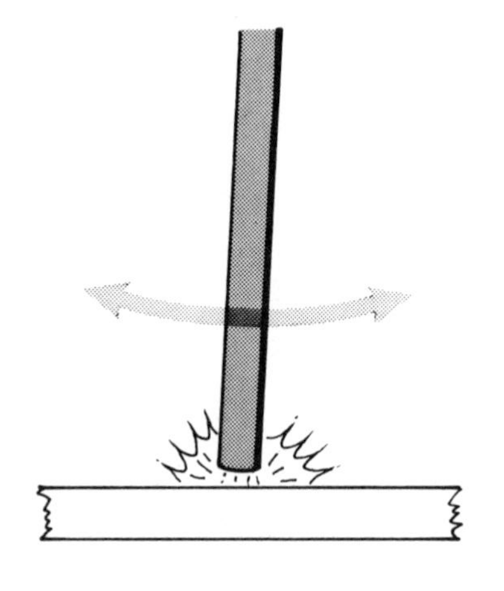

There are many ways to strike an arc, but with low-amperage home welders, and no experience, a sweeping or dragging move over the work surface may prove easiest.

As soon as the arc is struck you must prepare to move along and at the same time maintain the arc gap. Canting the rod in the direction of movement helps you observe the puddle. You'll learn to judge the arc gap by sound, but rod diameter is right for most jobs.

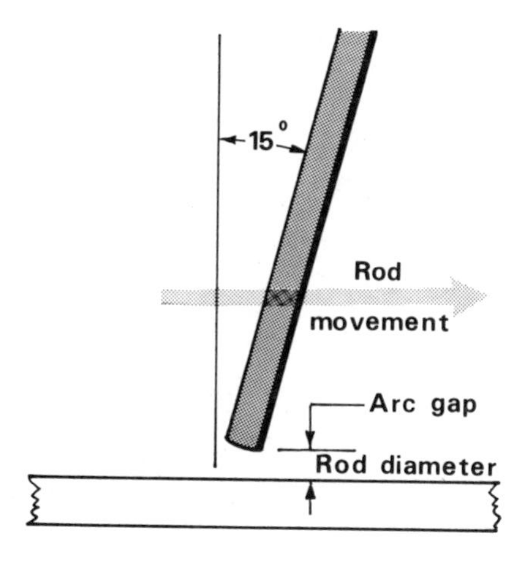

The next trick, of course, is to get the arc to burst into an established burn and then hold it steady and move it along to produce a bead. "Bead" is a term covering many things in all welding, but basically it means you've moved your weld along a path and deposited molten metal or fused metal behind. Many experts recommend holding the electrode holder with one hand and supporting it from underneath with the other. Steadiness is important, and the individual shakes of each hand tend to cancel each other out. In any case, be certain that you have a good, comfortable position and, if you're right-handed, with your left elbow tucked in snugly against your side.

Your first efforts should be directed toward running a continuous bead of metal, fairly straight, and of even width. At first you'll just get started and the arc will go out. You've overlooked the fact that as the electrode burns away the arc gap gets greater and eventually exceeds the arc's ability to jump the gap. Learn to move the electrode down at about the same rate as it burns. The ideal gap is about the same as the rod diameter, so ordinarily you will be trying to hold a gap of 1/16″ to 1/8″. Actually, nearly all experienced weldors judge the arc gap by sound rather than visually. A proper arc sounds like bacon crackling and sizzling in a pan. In a short time you will come to recognize the sound.

Watching the puddle. Once you've more or less managed to maintain the arc you must learn to concentrate on the puddle. Your electrode will normally be canted about 15° to 20° toward the direction of travel, and the puddle should be clearly visible. Try to watch the cooling metal just behind the puddle and to hold the bead width and amount of metal deposited constant. That's what welding is all about—depositing a good bead of metal without piling up or burning through the workpiece. The deposited metal will also guide your rate of movement. If you're moving too fast the bead will look like a narrow dribble of metal and won't be well merged

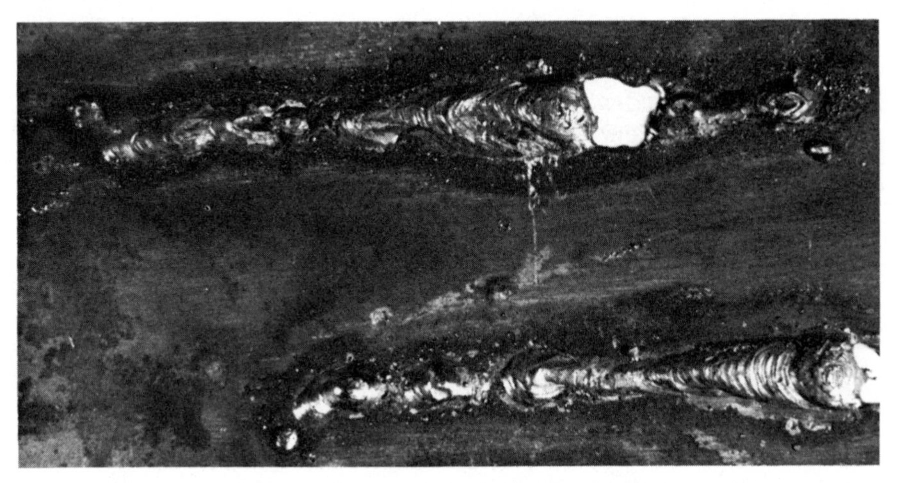

Don't be discouraged if your first beads look like this. Basically, welder started at left by moving too fast; didn't let puddle form. Metal dribbled onto work. He then slowed down to get a better puddle, went too slowly, and burned through.

into the work. If you move too slowly the bead will spread out too wide and start to sag through the work.

A right-handed person will normally move the electrode from left to right, because that's natural and makes it easy to see the bead trailing behind the puddle. If you're left-handed, weld from right to left. Later, you are certain to find circumstances that make it necessary to move the electrode in less orthodox ways.

It's perfectly possible to hold the electrode steady and do a good job. But, as you practice, you may find it easier to "weave" a bead by moving the electrode tip slightly from side to side. Or you may find that a slight circular motion works best for you. Practice all of these methods until you can strike an arc consistently, run a fairly even, straight bead, and control the deposit of metal.

What makes a good weld. Every weldor, beginner or pro, should be his own strongest critic. This is especially important in learning, when it's easy to develop bad habits which are hard to break later. One such habit, for example, is holding too big an electrode-to-work gap. Another is starting to move the arc too soon at the beginning of a weld before the puddle has really formed.

Right now, before you even try welded joints you should learn to control and inspect your work even though you're only running beads on flat metal. Take your soapstone marker and rule some horizontal lines about ¾″ apart on your workpiece. In most of your future welding it will be necessary to follow a line or a joint, and you should try to do so now. This is not easy. Examine the practice pieces of most beginners and you'll see that beads often start off in the right direction and drift to one side as they progress. But a good weld goes where you want it, so you'll have to try to watch the puddle and keep a spare eye on the line.

How your weld should look. A good weld will always look good. The metal you've deposited will appear clean, evenly rippled, and without holes and slag pockets after you chip away the slag. *Caution:* Be sure to wear protective glasses or a face shield for chipping slag. The fragments are thin and very sharp, and will bury themselves in your eye tissue. A

After running each practice bead, take time to examine your work critically. Ask yourself: Why does it look the way it does? What should I have done differently? What changes should I make next time?

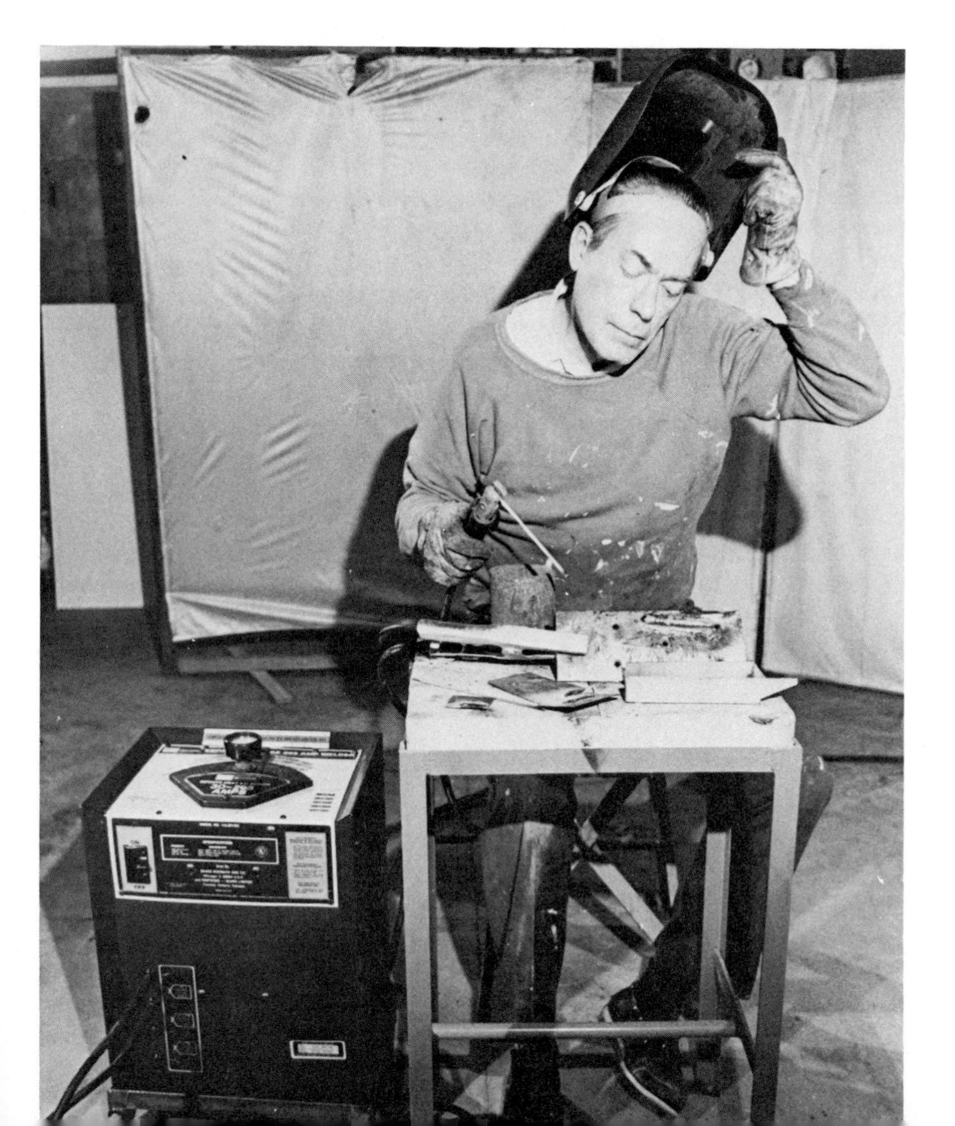

Lay out a practice piece with a soapstone marker so you'll have definite lines to follow. The purpose is to train your eyes and hands to make adjustments as automatic reflexes.

Your practice welds will be uneven—show improper rate of movement and failure to hold a constant arc gap. Wandering from the line is natural, because you're trying to manage so many things at the same time. Eventually you'll wonder why you had these problems.

good weld will be the same width over its entire length and its contours will not change.

Varying contours are often caused by using too high an amperage setting. The beginning weldor will strike an arc, not wait for a puddle to form, and start moving the rod. Thus the first part of the bead results primarily from melting metal from the electrode, not from mixing electrode metal and work metal in a puddle. Such beads have a "stuck-on," wormlike appearance. Then, as the work metal heats, the excess amperage causes the bead to widen out and maybe sag through the metal. At this point the weldor tries to correct by increasing the arc gap. Again, the appearance of the weld changes, this time to a spattered, ragged deposit. In the final analysis the basic error, other than moving too soon initially, was trying to compensate for excess amperage by adjusting electrode movement and gap. All of these problems will arise when you're learning, and the important thing is to concentrate on what you're doing, second by second, and analyze what went wrong when you examine your work.

Fusion and penetration. It's quite possible to produce welds that *look* very handsome but are not strong at all. Back in our early discussion where we used the analogy of two wet ice cubes freezing together I made the point that the water film actually merged with and became part of the cubes. In welding, of course, we want the metal of the electrode to merge or "fuse" with the work metal and become part of it. That's what weldors call penetration or fusion.

Right at this stage of practice, while you're still running beads on flat stock, you should learn to check your work for good fusion. If you were attending your local welding class you'd probably see your instructor pick up your practice piece and examine the reverse side from the weld. Unless you're practicing on exceptionally thick stock he'd be expecting to find a darkened area and evidence on the surface under the weld that your puddle actually penetrated well into the workpiece and, on thinner stock, almost came through. Obviously, too much penetration, burned-through holes, and metal droplets on the back are undesirable. As you practice, examine your work for real fusion, and don't be so concerned with appearance that you overlook this single most important feature of a good weld.

It's easy to pile up a bead on top of a joint without really penetrating and fusing the edges of the work fully. Low amperage and too large a rod may contribute to such bridging. Practice until you can fuse almost all the way through.

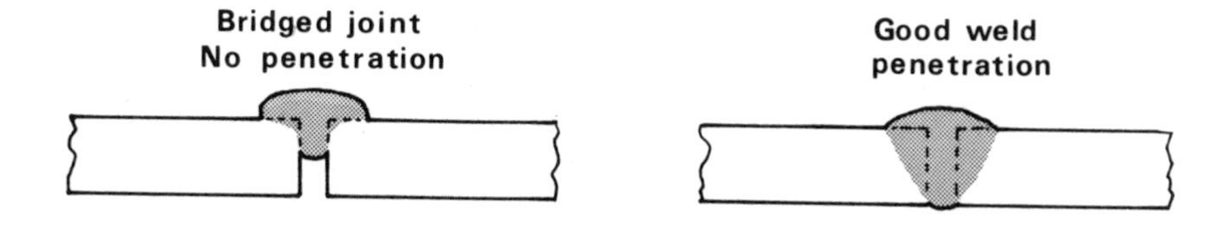

SERIOUS PRACTICE. Now that you've learned to strike and manage an arc, and have some idea of what a good weld should look like, it's very tempting to jump into trying various joints between two pieces of metal. I recommend holding off until you've done a bit more practicing, as follows. Repeat the sequence of ruling lines, following them with a bead, and inspecting and analyzing your work rigorously several dozen times, over and over. The object is to develop reflexes so you won't have to control each move consciously. Later you'll have enough to think about, and handling the arc must be fairly automatic.

So far you've been moving the electrode from left to right—at least you have if you're a northpaw. Southpaws go right to left. Not all welding jobs permit such convenience, and as you get into welding joints you'll find versatility handy. For a start, try ruling the lines vertically and starting at the top edge of your practice piece. Move the electrode and bead downward—toward you is perhaps more accurate. As before, cant the electrode 15° or 20° in the direction of motion. For a variation, start at the workpiece edge closest to you and work up or away from your body. After wearing out several practice pieces with vertical welds, try starting out on a horizontal path and altering it to vertical up or vertical down en route.

The movements above will feel quite uncomfortable at first. The object is to train your hand to cant and hold the electrode properly, and automatically, regardless of the direction of movement. A second object is to train your eye to watch and control the puddle even though you're seeing it from a different aspect. Later, when you set out to make a repair weld from an awkward angle and position, you'll find these abilities useful.

As a final word, I stress that welding is not a shared hobby. Especially during the early learning process, but, in general, always keep everyone, including wife and children, out and away. Arc flash can cause severe eye damage. Flying spatters can burn badly. And you don't necessarily want witnesses to your initial frustrations. Even if you're not embarrassed, the distraction can result in your absentmindedly picking up a red-hot piece of metal or doing something else foolish. Make it a firm rule. Always practice alone.

6 Welded Joints

From here on much of what you learn about joints and procedures will apply equally well to both arc-welded and gas-welded joints. The actual joints are much the same, and training your eye to watch the molten puddle is essentially the same. In designing welded structures, engineers will normally use certain basic joints to secure two or more edges together. Later, as you plan your own projects you'll also find your thinking following the same pattern, because these are the easiest and strongest joints and because they are accessible for welding. For example, if you were planning to bolt or rivet two pieces together edge to edge, you'd have to provide some sort of backup piece with holes in it to bridge the joint. As a weldor you'll think in terms of a simple butt joint. Or if you want to fasten two pieces at a right angle to each other with bolts or rivets you must somehow provide an angle piece in the corner to act as a joint maker. But with welding you simply fit the pieces together and run a bead or fillet in the corners. That's a tee joint. If two edges or corners come together, you can simply fuse them along the seam in an edge weld, or you can overlap two pieces, run a fillet at the edge of the lap, and have a lap weld.

The illustrations show these various classic joints. It would be nice if all your welding challenges were confined to such basic joints arranged so the work was always flat before you. In real life, however, and more often in home-shop situations than in industrial welding, you'll find it necessary to repair breaks, add reinforcements, and make structures which are not shaped so conveniently, are badly positioned, and are hard to reach. As with any skill, it's necessary to learn the scale before you can play the music. The practice joints explained in this chapter will help you.

A few more words about your practice pieces are in order. Try to pick up scrap from a shop that does punch-press work or stamping work with sheet steel ranging from 1/16″ to 3/16″ or even 1/4″ in thickness. Avoid

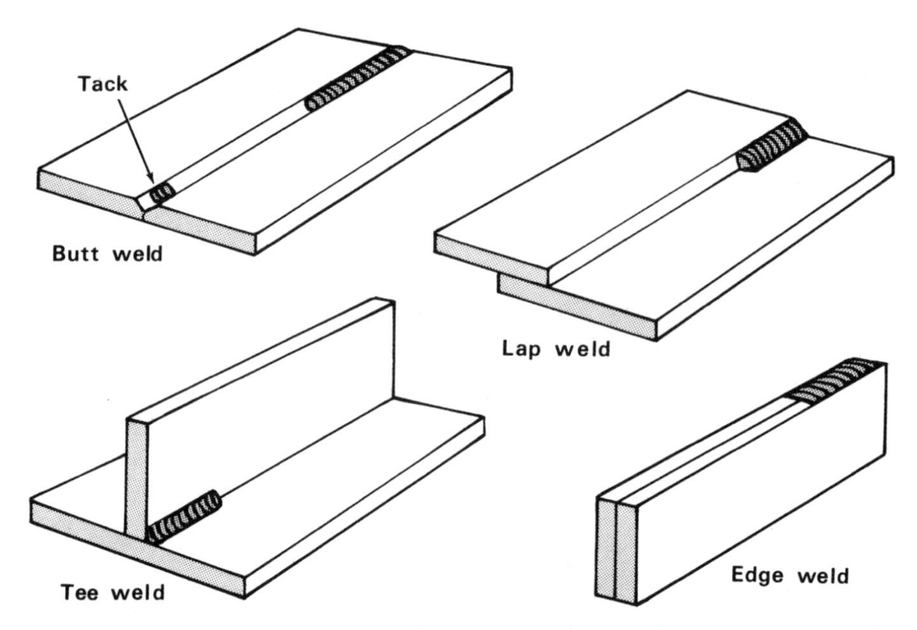

Classic welding joints in both gas and electric practice are the butt, lap, tee, and edge. You won't always find these in your work, but variations and combinations of them make welding challenging. These are the basic joints to practice first.

galvanized or cadmium-coated stock, because it not only will cause problems with your practice welds but also gives off heavy, white, toxic fumes. There will probably be times when you'll want to weld galvanized pipe or the like, but try to do it outdoors and avoid prolonged exposure.

If you read many welding instruction manuals you'll spot the fact that most suggest practicing on ¼″ or even ½″ plate. This is fine if you have heavy professional equipment and expect to enter industrial welding where such materials are routinely used. Actually, in 35 years of aircraft and home-shop welding I can recall welding only one piece of ½″ steel, that about 8″ square to provide a support on a basement H-column for a small arbor press. I've welded lots of angle iron and U-channel up to 4″ wide, but even so the metal thickness is seldom over 3/16″. More often, you'll find yourself welding light angle iron, ⅛″ or thinner plate, auto-body metal, light pipe and tubing, and the like. The techniques aren't significantly different, but welding heavier stock is easier, so why not start the hard way with lighter material such as you'll almost always be working with.

BUTT WELDS. Now's the time to have a go at actually joining two pieces of metal. This will be a butt weld and will be different from running beads on flat stock for two reasons. For one thing, you'll be placing the bead in a joint or gap. Secondly, you're going to encounter expansion and contraction, the universal bugaboo of all metal work. Your workpieces should again be $^{3}/_{32}$″ or $^{1}/_{8}$″ mild steel, and if you can find some strip stock $^{3}/_{4}$″ or 1″ wide and chop it into 5″ or 6″ lengths you're ready to go.

Place two such strips long edge to long edge on the welding table. It will help if you weight or clamp them at the right-hand end to keep them from sliding around. You can use C-clamps, although little spatters of weld metal will stick to the screw threads and eventually make them useless. Try to keep the edges about $^{1}/_{32}$″ apart. If this was a real-life, serious job you would probably want to grind or file a bevel on the edges to let the metal flow down between the pieces better, but for practice just leaving a small gap will do. Note that this is a difference between working with heavy stock and light stock. If you were going to butt-weld any metal from $^{3}/_{16}$″ thickness up, the bevel would be necessary.

Expansion and distortion. Start your weld at the left end and move along as you did on your flat plates. Just as all appears to be going well you'll probably notice some odd things happening. Your puddle will seem to be getting wider and sinking into the work. Slowing down the movement of the rod to add more metal doesn't seem to help. Stop welding and take a look. Ten to one the unwelded end of your practice piece has moved apart so the gap is wider and the puddle is more or less falling through the crack. Even worse, the two pieces which started out nice and straight have started to curl a bit. That's expansion and contraction at work. The heat caused the metal to expand and open up the gap, and the cooling weld metal contracted and added to the problem.

Tack welding. Later, we'll talk about how fit-up, clamping, and planning your weld procedure can help combat heat distortion, but for right now, and for most times in the future, your best answer is something called "tack welding." Try the butt weld again. This time place your clamp or weight temporarily in the center of the two pieces and make a short weld about $^{1}/_{4}$″ to $^{1}/_{2}$″ long near each end. These are called tacks and serve to hold the two pieces in the correct relationship during welding. If the

Starting at one end of a butt weld and heading for the other will almost certainly cause opening up of the joint because of expansion. Expansion is a powerful force and often overcomes clamping.

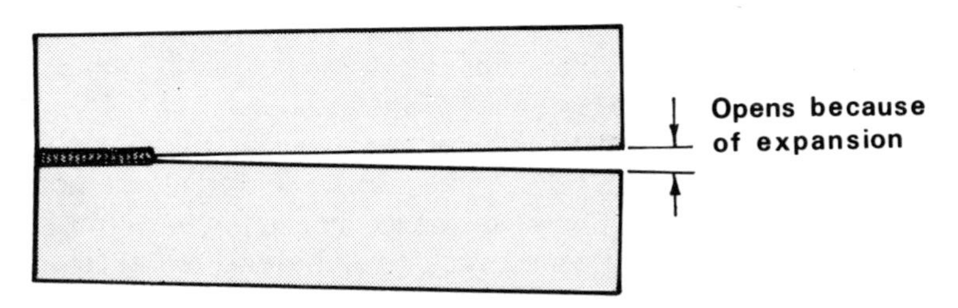

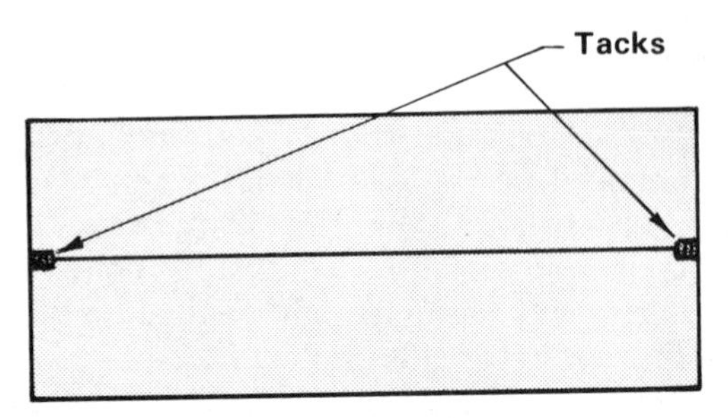

Tacks, short welds to maintain relationships the way you want them in a joint, are the weldor's best friend. The trick is to run them in strategic places. You'll go over them later, so don't try for full penetration and don't build up metal.

pieces were longer you'd want to place several tacks at about 2″ or 3″ spacings along the length of the joint.

As soon as the tacks cool, chip away the slag and proceed with the weld. Never start a new weld over slag.

Check your weld. Your butt weld should be examined for all the features of a good weld, and particularly for penetration. With a low-amperage welder it may be necessary to weld both sides of the joint, since the penetration will not go more than halfway through. This is especially true with 50-amp welders. If you feel your penetration was fairly good, try clamping your piece in a vise with the jaws just below the split line. Whack it hard with a hammer from the unwelded side. If your butt weld folds over, with the weld material serving as a sort of hinge, you didn't get penetration. One remedy would be beveling the edges. Another helpful technique is to use 1/16″ rod and be certain to get the puddle right down into the bottom of the beveled groove. After chipping it clean you may want to add a second pass. Such problems are typical of low-amperage welders and require a little extra forethought to conquer.

Depressed beads. Let's consider the opposite situation. Your weld looks good, but the bead appears concave and sunken below the top surfaces of the workpieces. This probably indicates good penetration, and the hammer-and-vise test probably won't show a hinge-effect bend. In fact, you may be able to bend the joint almost double and closed on itself without failure. About all that's wrong, if you want to be persnickety, is that you used a trifle too high an amperage, or too small a rod, and the thickness of the metal in the center of the weld is less than the thickness of the work metal. Hence if the joint was highly loaded and subjected to constant flexing, the stress would be concentrated in the weakest point right down the center. An analogy would be a piece of wire nicked at one point and flexed back and forth until it broke at the weakened point. In most home-shop welding this would not be critical, and admittedly few such edge-to-edge butt welds are used in practical construction.

Penetration to adequate depth in a butt weld may be difficult with low-amperage welders. Here, using 70 amps, only top surface has been welded. Weld shows "hinge" effect when struck with a hammer. Equivalent weld, below, using 100 amps, shows nearly full penetration and is adequate for most purposes.

LAP WELDS. Lapping one piece of metal over another is a fairly common repair procedure, and it has some of the same pitfalls as butt welding. In most cases, tacking is needed if the joint is more than a few inches long. It's also very easy to get the hinge effect if you don't get the puddle right down into the bottom of the edge notch. In fact, this happens more commonly with lap welds, because you're actually melting down the sharp upper edge of the lapped upper piece and the metal tends to roll down and cover the inner corner of the joint without fusing.

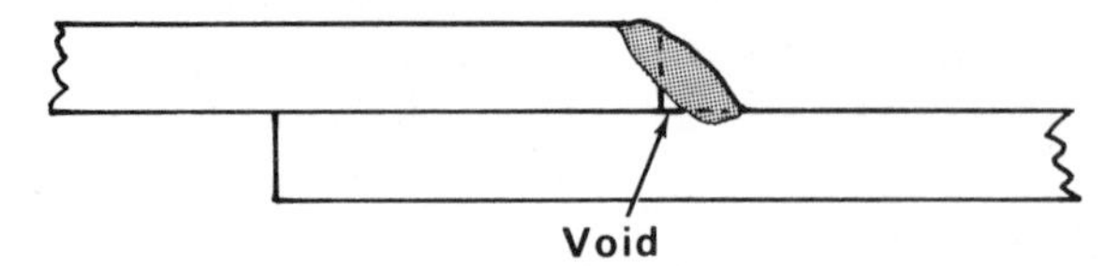

Sharp edges melt down easily and in the case of a lap weld will tend to overrun inner corner of a joint. This leaves an unwelded void, usually filled with slag, and often prevents good fusion into lower piece. Pieces are "stuck together" but joint is weak.

Directing the arc. At this point it's necessary to learn some of the rudiments of directing and controlling the arc heat. There are basically two methods you can use, often in combination. The first involves watching what is happening and adjusting the angle and direction of the arc so it heats one side of the weld slightly more than the other. The second is timing the slight moving action, weaving or circular, so the arc dwells slightly longer in one area of the weld than in the other.

In the case of the lap weld you should try to keep the arc and puddle down in the corner and away from the sharp upper edge by holding the rod tip down and controlling the puddle so it forms and runs at the base of the lap joint. When you have a good flow in the corner, a momentary weaving of the arc upward will melt the corner material so it flows in and fuses with the puddle. This technique and variations on it are used in many welding situations. As you practice, take time before you start the weld to visualize how the metal is likely to react, such as a sharp edge which wants to melt faster than the flat surface, and plan your electrode control accordingly.

EDGE AND CORNER WELDS. When two pieces of metal are flanged to bring two upward-projecting edges together, or when two corners are positioned to form a 90° outside corner, you have the beginner's dream for making a nice weld. True, tacking is necessary, but all you really want to do is melt the two edges so they run together. In gas welding it's easy to make such welds with no filler rod. In arc welding you will, of course, deposit metal. Light angle iron is handy for practicing. You simply tack two pieces flat to flat with the edges even. Practice edge welds with a little lower amperage and smaller rod than you'd use for butt welds. You'll find it's fun to "sew" two edges together.

FILLET WELDS. Although all of your practice so far has been directed toward welds where the metal has been positioned on the flat and the action is much like squeezing toothpaste from a tube, it's likely that when you actually start to build things with your welder you'll run more fillet welds than any other type. As mentioned earlier, a fillet is a blended deposit in a joint between two surfaces more or less at a right angle to one another. If you've ever placed a ribbon of caulking compound around a window or shower-stall joint you've probably created a fillet. Basically, a fillet is a wedge-shaped bead in a corner.

Your first fillet practice should be with flat scraps like those you've been using. This will be a long, straight fillet, and though not easy to make perfectly it is easier than fillets between two rounded sections such as you'd encounter in making a tee joint with pipe.

Holding your work. In all cases, flat stock or round, you'll have to devise a method of holding the two pieces in position. This can be a special problem for beginners, since most of us tend to stick rods and pull fit-ups apart when striking an arc. Later, we'll talk about fit-ups and clamping, but for practice it's often easiest to simply clamp one piece at right angles to another in a vise. If you don't want arc burns and spatter on the vise, drape it with wet asbestos paper. The important thing is to have your work solidly clamped so you can concentrate on the welding procedure.

Easy and hard fillets. Some welding instructors will start you on fillet welding by positioning the two pieces before you like an open trough. This is easier than having one face vertical and one face flat, since the fillet metal tends to distribute itself evenly on both faces. If you want to try this, clamp the pieces in the inner corner of a length of angle iron and support the iron so the 90° inner angle is evenly positioned upward. This is the ideal way to run a fillet, and whenever you can position your work this way it's so much the better.

The trouble is that you'll almost never encounter such an ideal positioning of the work on real jobs. Go ahead and try a few fillets this way, but soon you'll get the feel for watching both sides of the 90° joint for even

When practicing fillet welds, it helps to eliminate the need for tacks by clamping pieces solidly in vise. This is a harder fillet to run because you have both vertical and horizontal surfaces on which bead must be deposited evenly.

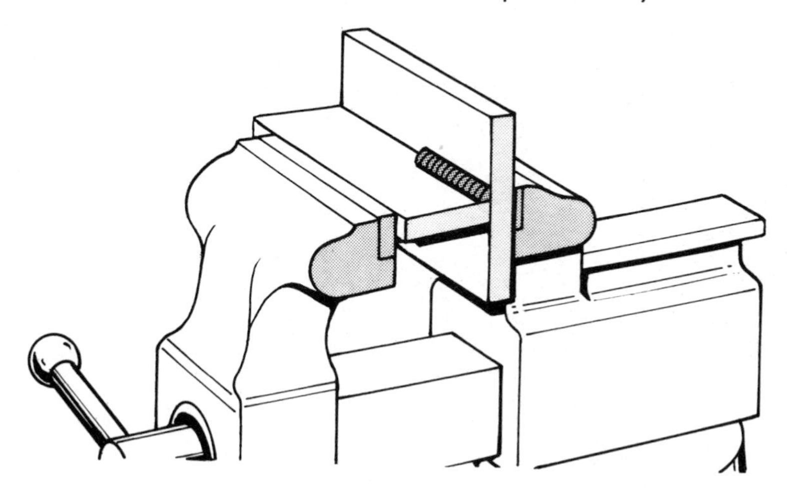

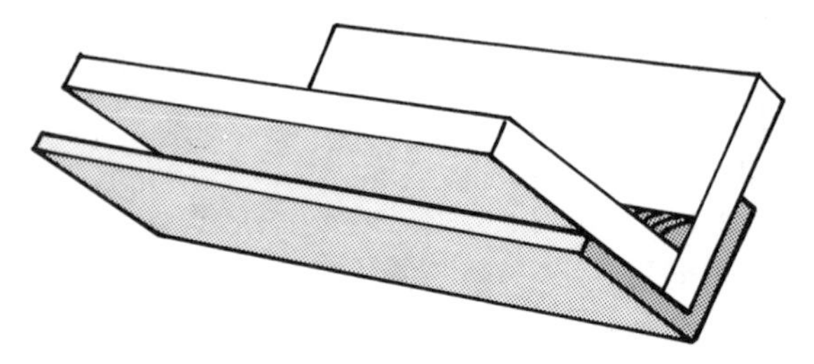

Easier way to practice fillets is to clamp workpieces in a trough of angle iron so joint is facing up and both sides receive metal deposit evenly.

bead distribution and it's time to try the "hard way" with one face vertical and the other flat. You'll find that the fillet metal has a mind of its own and wants to either run down and pile up on the flat surface or hang up on the vertical surface. This is when it really becomes necessary to practice directing the torch heat and timing the length the heat dwells on each face so you get an even bead.

Undercutting. Just about the time you feel you've mastered the touch of keeping the fillet up on the vertical face you'll probably notice a new problem called undercutting. The metal has actually slumped away from the face and left a depressed groove all along the upper edge of the bead. Too much heat on the vertical face, either because of the angle of the electrode or your movement of it, or perhaps too high an amperage, can

This practice fillet weld started off well, and actually looked fairly good until slag was cleared. Filler metal didn't get into corner. Result is no weld at all.

Voids and slag pockets in the corners are even more common with fillet welds than with edge welds. As you practice, saw your welds in half, or clamp them in a vise and hammer them to see how good they are.

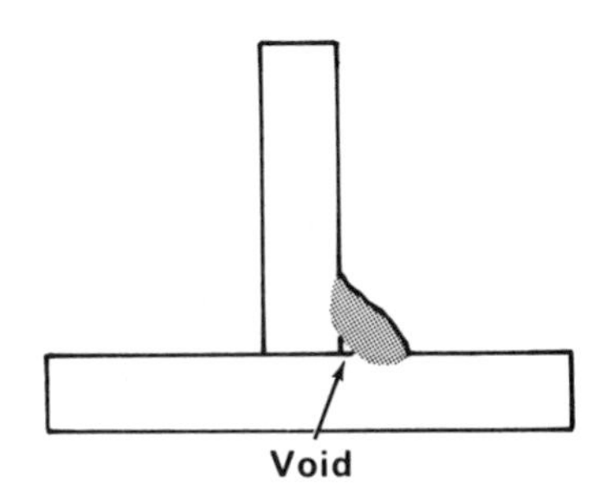

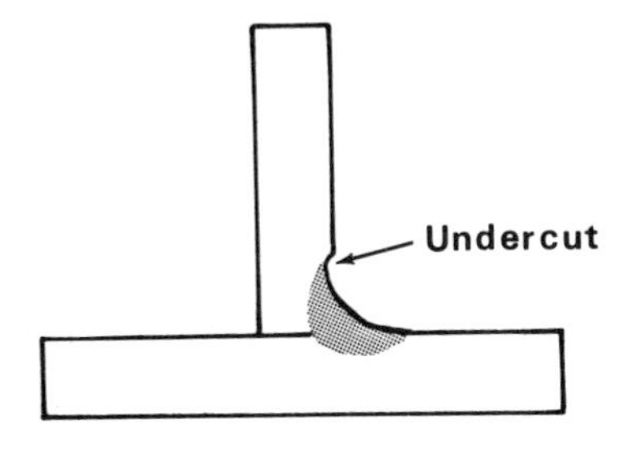

The usual result of solving the problem of a corner void is a new problem, undercutting, which means a depressed edge along the bead. Unless undercut is severe, this is preferable to a void if you have good fusion.

cause this condition. Again, practice, careful observation of what you're doing and what the metal is doing, and trying little changes in technique are the stuff from which learning is made. Your own self-criticism is as important as any guidance an instructor can give you.

Fillet penetration. Since fillet welding is so much a part of building almost anything by welding, you must devote more time to practicing this type of joint than to the simpler joints. If you recall the hinge effect I described when talking about butt welding, you'll remember that it was caused by failure to get the metal down into the bottom of the joint between the two pieces. Some of the suggested remedies were spacing the pieces slightly apart, beveling the edges almost to the bottom of the joint, and using a smaller rod to concentrate the heat in the bottom of the groove.

This same failure to penetrate down into the root or corner is even more common with fillet joints. You may find that the fillet metal wants to "drape" itself in a tentlike fashion from the vertical face to the flat face

Ideally, your fillet welds will have no voids or undercuts and the fusion will penetrate deeply into both pieces. In home-shop welding you may never penetrate fully to opposite side. If full strength is important, run another fillet on reverse side.

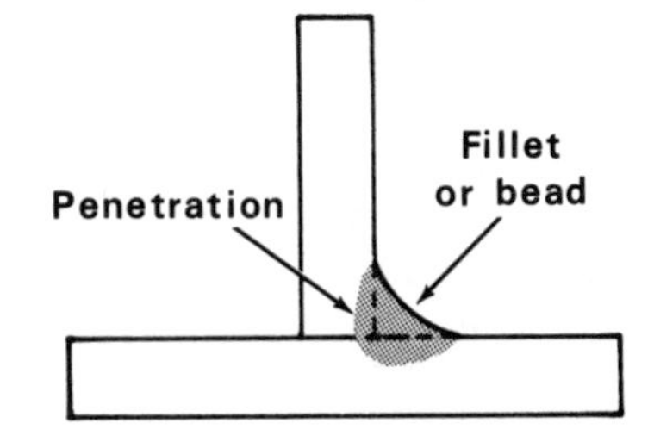

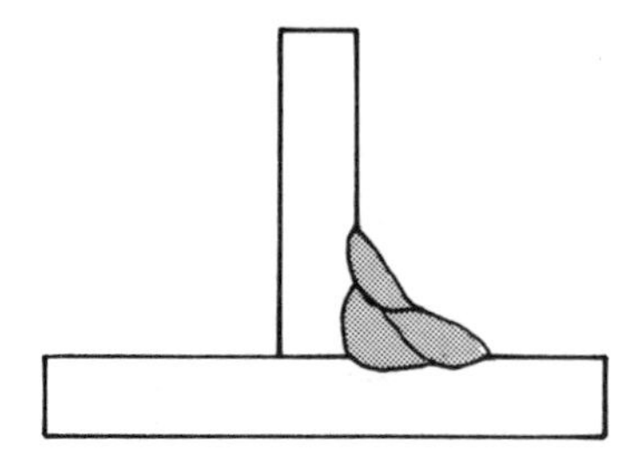

Multiple fillets, starting first with a small rod right down in the corner and building later with successive passes to provide a full joint, are an accepted technique in industry.

and leave an inner channel or groove with no weld metal at all down in the corner. This is particularly common with very low-amperage welders. A common cause is using too large a diameter rod, which tends to bridge the arc across the two faces. This also tends to trap gas and slag down in the same troublesome corner. Remember, I'm not writing this for industrial or professional weldors. They have ways to handle such situations including choices of equipment, rods, and polarity not commonly available in the home shop.

Successive beads. The easiest way to get full penetration into the corner of a joint is to use small rod, 1⁄16″ for example, and work the puddle right down at the bottom of the corner. This may not give you the fillet width you want for a strong joint. The answer is to chip and wire-brush away all the slag and run a second bead along the lower edge of the first. Again, use a small rod. Repeat the cleaning and place a third bead, this time along the upper edge. Each of the last two beads should overlap the first, and their edges should merge over the center of the first. If desired, even a fourth bead can be run down the middle of the fillet, or you can step up to a larger rod and run a finish or "dressing" fillet the combined width of the others.

Different metal thicknesses. So far your fillet welds have been on flat, straight surfaces, but your real-life welds will not always be so nicely set up. Nor will the metal pieces always be of the same thickness. Try coping with the latter problem first.

Almost any combination of thick and thin metal will do, but for a start try placing a piece of 1⁄16″ steel sheet on edge to form a 90° joint with a 3⁄32″ or 1⁄8″ piece. Now you're going to have to use enough amperage and hold your puddle long enough to penetrate the heavier stock but avoid burning through the light piece. Most experienced weldors will manage this by the weaving technique I've mentioned before. The motions will vary from minute side-to-side oscillations of the electrode tip to a small oval pattern in which the arc passes the lighter stock quickly and drops down to move more slowly over the heavier material for slightly greater heating. Different weldors develop their own styles, but the fundamentals remain the same—control the amount of heat directed to the thick and

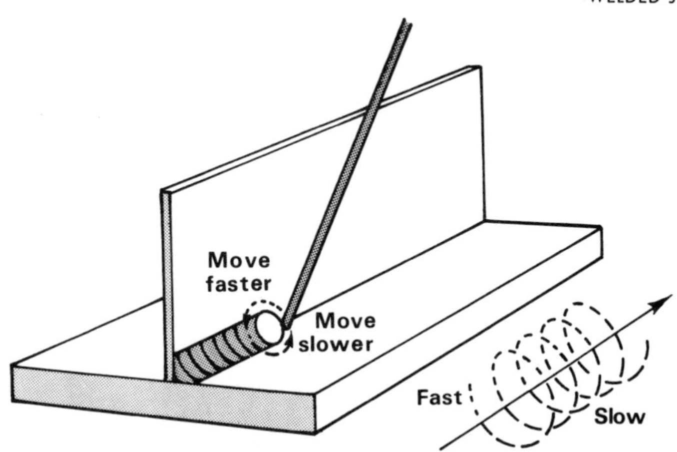

Many times you'll find that more heat is needed in one part of a joint than another. Example is welding thin metal to thick. Here, the electrode tip is circled slightly and arc is held longer on thicker stock than on thin.

thin stock by timing the arc movement as it is alternately directed toward first one surface and then the other.

There are whole chapters in professional welding books on weaving techniques and tricky little movements to place the right amount of metal and heat where you want it. This is part of the "art" of welding and like an artist's brushstrokes has much to do with the quality and appearance of the finished job. This artful touch is important, but only practice will refine it for you. Don't be afraid to experiment.

FILLING POOR FITS. One such experiment you should make is welding together two pieces with edges which touch at only a few points. Such situations are common in repair welding, even though it's always better to have snug fit-ups. You make some practice pieces by grinding an edge on one to an irregular shape which will leave gaps of ⅛″ or more at some places when it's butted beside another piece. Obviously, if you simply try to run the arc along such a joint the metal will fill where the edges touch or approximate and will fall or blow through the larger gaps. Here, the trick is to watch as this tendency develops and move the arc puddle back a trifle to build a little "glacier" of cooling metal to bridge the gap. Now advance the arc and deposit metal along the leading edge of this glacier and again back up and allow a fraction of a second for cooling. You'll find your rod tip moving in a sort of horseshoe or U pattern with the legs of the U extending in the direction of movement and slightly ahead of the advancing metal. Basically, you're building ahead on the leading edge of your deposited metal.

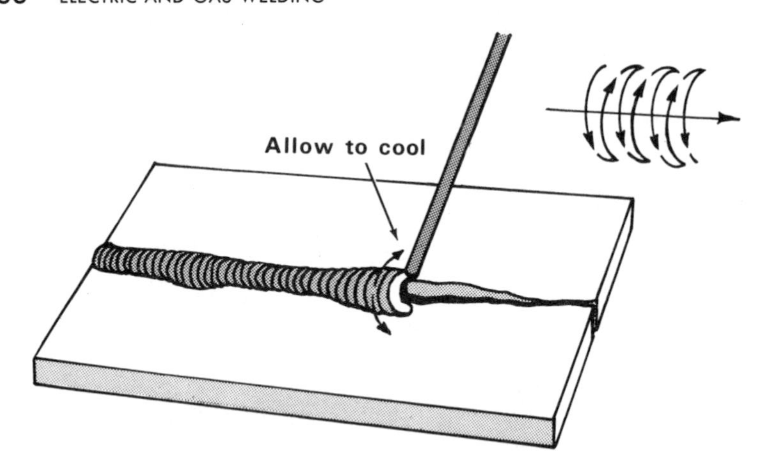

Time your arc and control its pattern when welding joints that don't fit perfectly at all points. Gap in butt weld shown here can be handled by moving arc in a U-shaped path and giving metal in center of path a slight chance to cool so it doesn't remain too fluid and drop through gap.

All of the above exercises will sharpen your eye for the position of the puddle, the angle and relationship of the rod to it, and the appearance or failure to appear of good fusion. These things are important, because you want to make this conditioned-reflex response to the appearance of the puddle an automatic guide for your hand.

NON-FLAT JOINTS. You'll discover what I'm talking about when you start welding joints which are not flat and straight. For a start, saw off a short piece of 1″ or 1¼″ pipe. The ordinary black steel variety will do, and all you need is an inch or two. Place a piece of 3⁄32″ or ⅛″ flat steel on the welding bench and up-end the pipe on it. You can rest a piece of firebrick or heavy metal on top of the pipe to hold it for tacking.

Your objective will be to weld a fillet all around the pipe at the joint with the flat stock. Before you start, try positioning your body and, without turning the welder on, run the electrode around the weld to get the feel of the problem. One position, of course, would be to stand over the work and look down. This does not offer the best view of the joint but it might be the only position possible on some real jobs. Another posture might be to sit down and edge yourself around the work as you progress. This introduces the need to drag the rod-holder cord with you while trying not to break the arc or stick the rod as you shift position. Or, if you go in for production-type set-ups, you might want to cheat a little and make a small turntable so you can slowly rotate the work as you weld. I suggest trying all of these, since in many instances you will find both awkward postures and body shifts are required because of the shape of the joint. Most begin-

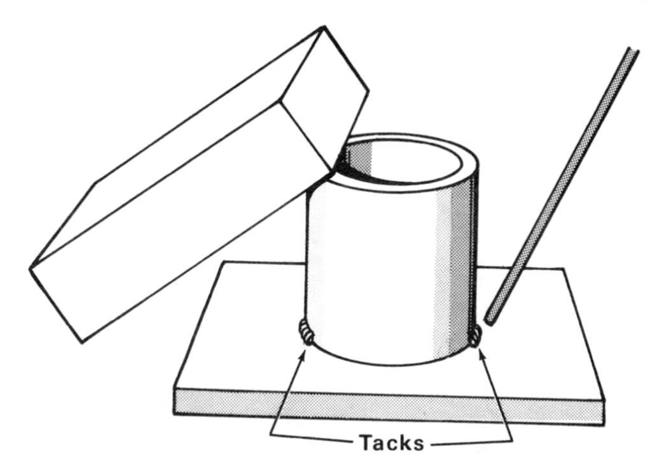

You need to develop automatic reflexes for constant readjustment of rod angle as you move around curved joints. There are many ways to position yourself and work around this practice joint. Start by tacking weighted piece of pipe. Then try experimenting with different ways to run a fillet around the base.

ners will end up by welding as far around the pipe as they can comfortably and then breaking off, chipping slag, and moving either themselves or the work.

Controlling rod angle. Whatever position you assume you'll find that you must combine not only the usual forward motion and, perhaps, weaving of the rod, but also a constant angular adjustment of the rod as you circle the joint. Practice this type of closed-end weld between flat and tubular stock until you can produce a clean, even bead all around. One way to check your work is to pour a little water or penetrating oil into the open ends of your finished practice pieces. If you leave them overnight and come back to find water or oil oozing out along the bead, you'll know you didn't get as perfect a job of fusion as you should. The joint might hold perfectly well structurally but it is not pressure- or liquid-tight.

ANGLED JOINTS. When you're satisfied by your progress with an open cylinder at 90° to flat stock, try cutting the pipe at 45°, more or less, so it will rest on the flat stock at an angle. Now you've introduced a whole new set of problems. In addition to welding around a circle with poor access and visibility on the side toward which the pipe leans, you've got to make a less than 90° fillet on one side, more than 90° on the opposite side, and are faced with constantly varying metal thickness because of the slanted cut across the pipe. To understand what I'm going to bring up next, go ahead and make a few such angled welds.

Beveling the edges. What I hope will be apparent from the above practice is that many welds are made better and more easily after beveling the

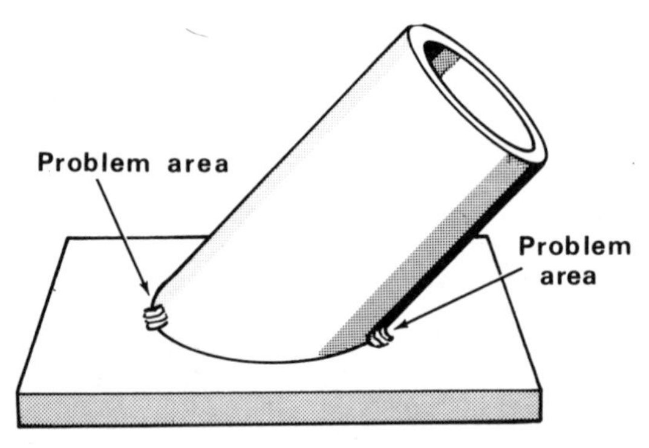

Angle joint introduces new problems of thin metal in open part of angle, and close quarters and poor visibility in acute angle. Again, practice and experiment are the keys to developing proficiency.

work edges. This, of course, is standard practice with heavy industrial plate, but it's not always needed with lighter materials. In the case of the angled pipe, however, try using a file or grinder to bevel back the thin edge at the acute angle of the pipe until you've got about a 60° open angle to run the bead into. Taper such a bevel about halfway around each side of the pipe and you'll discover it's much easier to do a good fillet weld.

FISHMOUTH JOINTS. By now you should be accomplished enough to try a still more difficult weld called a fishmouth joint. Such joints are commonly used in tubular steel aircraft construction. You are probably more likely to use it in building such things as playground swings, boat trailers, and the like from pipe and tubing. Even ordinary steel conduit can be used for a great variety of lawn, garden, and shop articles such as tables, shelves, and benches.

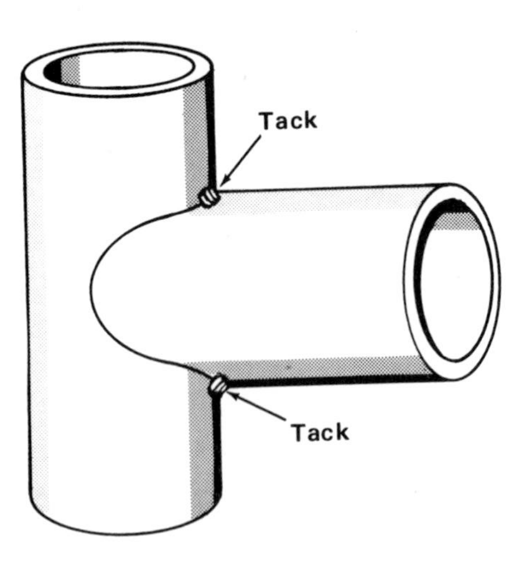

You can devise a variety of position problems with this simple fishmouth practice project. Tack it first, then try it over and over in tee-down, flat-on-side, tee-right or tee-left, and tee-top positions. Real weld jobs tend to present such position problems.

Rough tests of practice welds show that butt joint in flat steel bends without breaking and although not perfect is probably satisfactory for most home projects. Fillet weld with poor penetration shows "hinge" when struck with a hammer. Edge weld is rough, but sawing through shows that edges are well fused over top of joint.

A fishmouth is needed to join two tubular pieces in a tee or angle joint. Obviously a square-cut end will not join well to the rounded side of the mating piece, so the metal must be shaped in the form of two extended lips which reach around the sides of the rounded surface and are called a fishmouth. We'll talk about the shaping process later under cutting and fitting, but suffice it to say that you want a shape that intersects the curved surface neatly and without gaps and at the selected angle.

You should practice such welds at first by placing the pieces flat on the welding bench and tacking both sides. Or you can arrange one piece vertically to form an inverted tee. This is a type of weld where you can learn to exercise your own ingenuity. It has built-in problems. Not only must you follow around a curve and adjust the rod angle as you go, you also will encounter a transition from horizontal to vertical welding. If you invert the tee you can go from vertical to horizontal overhead welding.

The easy way, of course, on a small practice piece is to weld a short distance and then stop and move the piece so you're always welding flat down. With small subassemblies that's perfectly good technique, but if your fishmouth joint is part of a structure which can't be rolled and tumbled for position you'll have to learn to weld uphill, downhill, and overhead. We'll talk about that in Chapter 9.

TESTING WELDS. Welding literature is full of weld inspection methods ranging from bending, stretching in tensile machines, and deliberate breaking by fatigue to X-rays and radioactive photos to reveal hidden flaws, porosity, and slag inclusions. Few of us at home can resort to these techniques. It is a good idea, however, to take as many of your practice pieces as practical and bend them in a vise or try chiseling them apart. Another good learning experience is to saw through the weld crossways at several points. This will reveal failure to fuse the bottom of a fillet or lap weld and may show slag pockets you didn't realize were there. It is not unusual for a weld to appear good simply because the slag has filled in where metal should be. If you have access to a sandblast machine, by all means try blasting a few of your practice pieces. You may be surprised at what thorough slag removal reveals. Another revealing experience can be gained by buying a can of the spray-type dye penetrant used for industrial inspection. It will show up hairline cracks caused by shrinkage, too rapid cooling, and just poor fusion.

7 Fit-up: Cutting Your Workpieces

Industrial weldors are lucky. Their workpieces are almost always stamped, cut, or shaped so they fit together just right. And they usually have production jigs that clamp the pieces exactly in position. Often the jigs can be rotated for easy welding positions.

A home-shop weldor is seldom so blessed. His work runs from patching the locking lugs on his wife's purse to repairing a cracked joint in a bicycle or the angle-iron frame for a fireplace smoke damper. Often on repair jobs and always on new projects there must be fresh metal cut and shaped. Not the least of home-shop welding challenges is the use of material that was never intended for such use.

This has nothing to do with actual welding technology but will have a profound real-life application to using your welding skills. Acquiring a welder may even give rise to domestic conflicts. After a few searches through the dark corners of your shop for just the right piece of angle, bar stock, or plate, your eyes will automatically become alerted to the value of "junk." A discarded floor lamp in a friend's garage stands revealed to you as a valuable piece of square stock if you merely saw off the base and socket end. Discarded steel snow-fence posts make dandy uprights and supports for workshop storage racks. Your acquisitions, and likely the family protests over junk storage, will be endless. Personally, I always tuck away a useful-looking piece of metal. Sooner or later it will be useful in reality.

Even when you have the raw stock for your repairs and projects you can do nothing without some way to cut and shape the individual pieces. A hand hacksaw with a good fresh blade will get the job done, but even a simple frame for a welding table requires sawing quite a few pieces of 1¼″ angle iron. Larger pieces of round stock and channel can be tedious. This work may not overwhelm those with husky arms, but lesser souls or those with arthritis can find it more than they want.

A series of holes drilled with an electric drill or drill press, followed by a hacksaw, is a crude but work-saving way to cut heavy stock.

One solution is to locate a friendly local metal job shop with power saws and heavy shears. You can then lay everything out carefully, even to the extent of taping together a cardboard mockup, and hire the shop to cut them to size. The problem here is that after the first piece or two is welded you may want to change your mind or that a little heat distortion has made your fits "no fits."

Another possible procedure is to use an electric drill or drill press to do most of the cutting by drilling a series of closely spaced holes and following

Best shop tool for cutting metal is a bandsaw with a speed reducer and metal-cutting blade. Cuts are smooth and can be angled or curved. Electric saber saw can sometimes be used on sheet stock.

with the hacksaw. This is slow but less work. The redeeming feature of welding is that such rather messy edge cuts get filled with metal and no one knows the difference.

The bandsaw. The best solution in the home shop is a slow-speed converter for an ordinary woodworking bandsaw. The converter slows the blade down to the more or less correct rate for metal, and the metal-cutting blade does the rest. In addition to the obvious gains in speed and reduced elbow grease, the bandsaw permits curves and rounded contours you can't make with a hacksaw. Moreover, by using a miter guide you can cut accurate angles or neat square ends.

Bandsaw blades come in various tooth spacings for different metal thicknesses. You don't want coarse 8- or 10-tooth-per-inch blades for sawing thin sheet, because they snag and tear. And you don't want a fine 18- or 20-tooth blade for a heavy chunk of bar stock; it's too slow. For practical purposes a 14- or 16-tooth blade handles most ordinary work and will cut steel, cast iron, brass, aluminum, or almost anything except hardened tool steel. Such bandsaw blades are not expensive—they're not really cheap, either, but they'll last a long time unless you run them into something really hard. That something can be a localized spot in work you've welded previously and now want to saw apart. Some types of low-grade angle and strap iron are rolled, I'm told, from scrap containing mixed alloys. In any case, welding such metals can leave glass-hard inclusions that ruin a blade almost instantly. Before sawing into an old weld, test it with a file or hand hacksaw. If it feels glassy, forget about sawing it. Use a grinder to remove the old weld if necessary.

Bench grinder is indispensable in metalworking. For welding, it is the best way to bevel edges to get penetration.

The grinder. A good bench grinder is an important part of any shop, and especially so when you're welding. The uses are too many than to do more than suggest. Beveling and fitting pieces comes to mind first, but others include grinding off rivet heads to salvage old stock, sharpening drills and chisels, and cleaning away old welds, damaged areas, and scale. A stiff wire brush mounted on your grinder is great for slag removal. If you fabricate your own special tools—and that's one of the advantages of welding—you'll want to fuse your welds down deep into bevel grooves and then grind them flush so they're invisible. This also applies to crafts and repairs where welds would be unsightly even if painted over.

A second type of grinder you'll want to try to acquire, even used, if necessary, is a hand-held disc grinder of the type used in auto-body shops. This powerful grinder is extremely handy for cleaning up old rusted metal, but its greatest value is for cleaning up welds, for example on auto bodies, fabricated structures, or wherever a typical weld bead and associated spatter are messy-looking. A substitute is an abrasive disc in an electric drill or on a flexible-shaft drive.

Still other abrasive operations are part of putting together a neat weldment. Even bandsaw-cut parts tend to be ragged along the cut edges. Exposed edges should be dressed up before welding. My own experience

Nothing serves better for grinding weld beads flush, cleaning up metal, and removing spatter than powerful hand-held Milwaukee disc grinder. Abrasive disc in an electric drill is a substitute but less effective. Always wear eye protection when grinding.

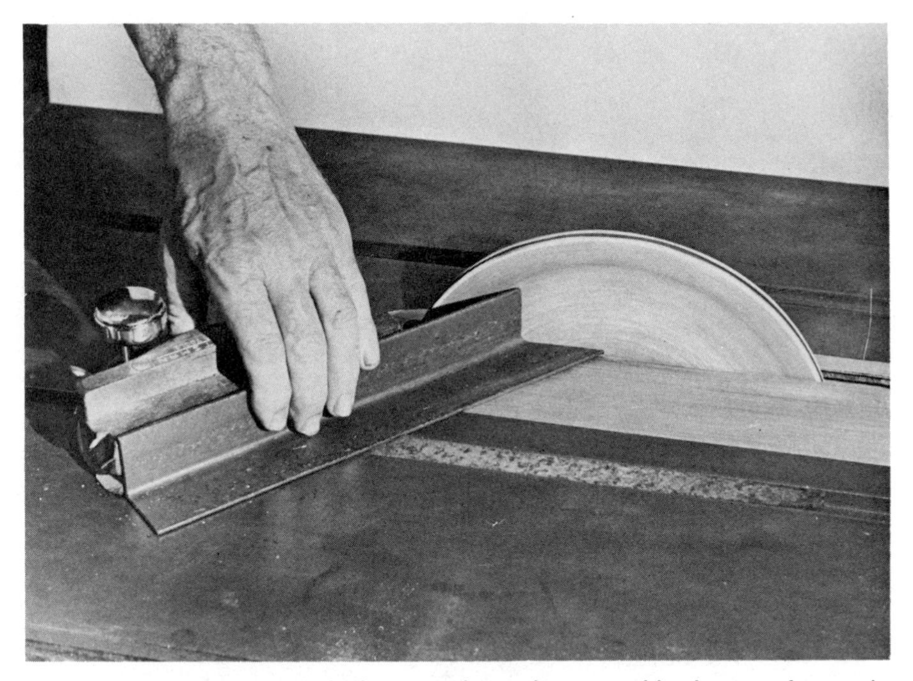

Accurate joint fitting, removal of saw marks, and preassembly cleanup of ragged edges can be done quickly and neatly on a table saw fitted with a metal-backed abrasive disc. Belt sander will also do a good job.

is that a 10″ metal sanding disc with self-adhering abrasive discs is about the best there is for such work. I mount the disc in my table saw, but if you have a belt or disc sander that's just as good. An example of how such an abrasive disc can help is the welding table shown on page 168. The four sides of the tabletop to hold the firebrick are welded together at 90° angles after first cutting the 45° miter joints on the ends of the 1¼″ angle stock. Even if your sawing is less than perfect it's easy to set the table saw or sander guide at 45° and slide the ends against the abrasive to trim the surfaces to a perfect miter fit.

Another example of the importance of grinding equipment is the wrought-iron table shown on page 7. The vertical decorative grill members must each be shaped and fitted to length to match the curved contours of the outer frame. This was easy with the grinding disc and table-saw guide.

Fishmouthing. As mentioned earlier, you'll often want to join pipe or tubing at right angles or other angles with a fishmouth weld. Such a weld requires a good fit. There are many ways of making the proper contour, ranging from ideal to crude. Ideally, you can have a hardened steel block with a hole to receive your workpiece and another hole drilled to intersect

RAPID

Start of fishmouth in steel conduit is a hacksawed vee cut, top left. With this type of joint, conduit is a very useful material and is easily welded or brazed into light, strong structures. Use aircraft tubing for more critical pieces. Second picture shows working rounded contour from initial vee with a half-round file. This takes practice before you can do it quickly and accurately. Third picture shows a rotary file in a drill press contouring fishmouth. An electric drill or rotary tool can also be used, but maintaining alignment is harder.

the first at exactly the angle you want. By inserting and clamping your work and feeding in a milling cutter, a hole saw, or even a drill through the intersecting hole you can produce a neatly fishmouthed contour which will snug up against the mating surface without gaps and at the desired angle.

This method is fine for production, where the same cuts must be repeated many times. But few home shops have large drill bits of exactly the right diameter to make the holes. And the time spent making such a tool is wasteful unless you have many joints to fit up. Nevertheless, if you're into a project needing quite a few such fishmouths it may be worthwhile to make the tool from hard maple or oak, using an auger or expansion bit for drilling.

Another, more practical approach is to approximate the open fishmouth with a bandsaw or hand hacksaw. Usually about all you can do is make a vee cut. This initial cut can then be rounded out with a rattail or half-round file and maybe finished with a rotary file or grinding stone in an electric drill, flexible shaft, or high-speed rotary tool. My favorite method is to fix a rotary file in the drill-press chuck, clamp the tubing in a drill-press vise, and work the shape out by sliding the vise on the table and just touching the walls of the tubing as needed.

8 Clamping and Set-up

As mentioned earlier, sticking an electrode and having to break it loose with a quick snap may not be a problem with large heavy work, but with lighter, more delicate assemblies you may face a real challenge in supporting and holding the pieces in place until you at least get them tacked.

If you browse through hardware stores and welding supply shops you'll find a great many ingenious clamps, holding devices, and alignment tools to ease the common clamping problems of commercial weldors. Some of them, such as the self-clamping plierlike tools, spring clamps, and chain locks, are extremely handy. In many cases, however, it seems you have to use your own ingenuity to cobble together some sort of holding rig.

My own collection of holding techniques runs about like this:

- Weights, ranging from simple firebricks to heavy pieces of scrap metal. By resting one end on the work and propping up the other end as needed you can often get by.
- Magnets, mostly small but powerful 3″ or 4″ bar magnets used as temporary "glue" to hold pieces for tacking.
- C-clamps, the almost indispensable clamps, although they will be ruined eventually by welding spatter.
- Cabinetmaker's clamps, the kind that clamp onto pieces of 1″ pipe to make a clamp of any length. These are great for holding together longer pieces of angle, channel, pipe, and like frames.
- Specially made jigs of plywood and chipboard made for accurate forming and shaping of duplicate parts. The Early American table ends shown on page 7 are good examples. Such jigs will, of course, burn and char some, but if kept well soaked with water while welding they'll last quite well.

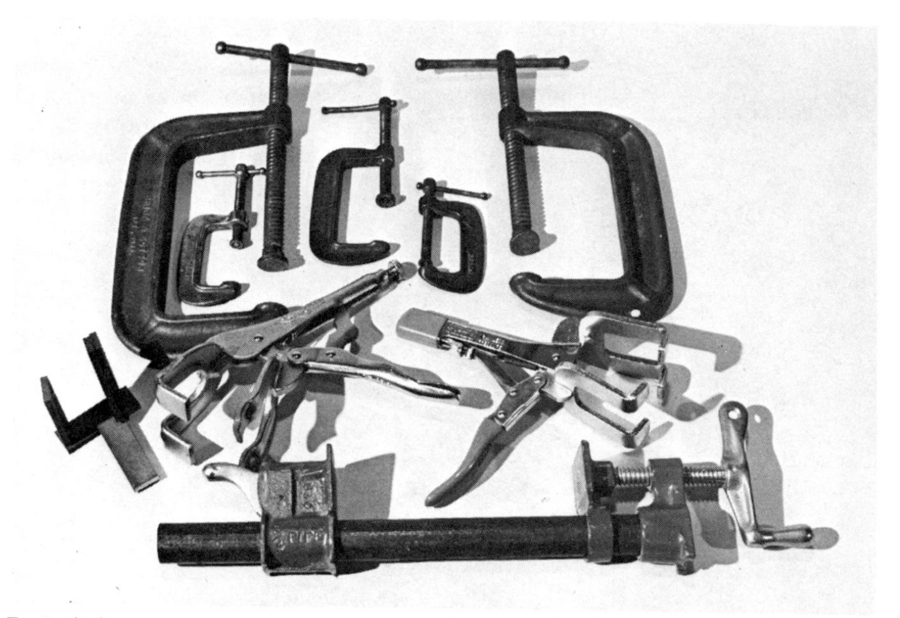

Typical clamps you'll find useful in holding your work for welding are large and small C-clamps, toggle-action clamps made especially for welding, and cabinet-maker's clamps for larger frame structures. Small metal squares, left, are bar magnets; they're great for holding small pieces for tacking, and also for securing shielding.

Whatever clamping or holding means you use for any given job, the most important thing is your own judgment. If the job appears a bit tricky, take time before you strike the arc to ask yourself what's going to happen when the heat is applied. What will happen as the weld cools and contracts? Is it better to do the job with a series of short welds spaced here and there with cooling time allowed between them? Should you work first from one side and then the other to even out warpage? How long a bead will you want to run before breaking off to see what's happening? Will your clamping pressure cause the metal to flex or distort when it gets red-hot? Will it be necessary, or possible, to bend and realign the assembly while it's still hot?

Bending metal. If bending for form or alignment is going to be needed, your work must be firmly secured so you can apply pressure. Ask yourself if such force is best applied with hammer blows or by gripping or slipping a pipe or something else over the work and tugging gently. Whatever means you elect to do the bending it's important to make a dry run to be sure you have the tools within easy reach, the jaws adjusted to fit in the case of a wrench, and a place to lay down or hang your torch. In short, when you flip up your helmet and set about bending you want everything

Although made of particle board, this jig for welding ends of theme table worked well, is still usable. Round cutouts provide clearance for welding. Fixture was kept wet while welding.

ready so the metal doesn't cool while you scramble for tools. Actually, although such things are sometimes called "skill," they're really common sense. In welding such planning and forethought is doubly important, because once a weld has been made it's very hard to change it and it may be necessary to start all over. If you try to program your mental processes you'll find that eventually your subconscious mind will automatically run through the questions I've mentioned and stab you with little hunches to guide you.

Keeping the heat where you want it. Arc welding has a great advantage in its ability to produce instant, intense, localized heat. In some circumstances, such as welding cast iron, this is a disadvantage because the surrounding metal remains cool and the strain between hot and cold produces a crack.

For the most part, however, the quick and localized nature of the arc is a great saver of time and parts. As an example, a neighbor who builds and races cars had a need for two small welds to secure components on the front axle of a car he was scheduled to race a few hours later. About all he had time for was to pull up to my garage door. The job could have been done with gas welding, but the time required to get the relatively heavy parts up to puddle heat would have allowed the heat to creep out to the wheel bearings, caliper brakes, and steering joints. The arc, on the other hand, quickly secured the parts in place and the surrounding parts weren't even warm to the touch. It took only a few seconds on each side.

I relate this because there are several things to be learned from the experience. He didn't learn them because I didn't mention them, but for your information he committed two important "no-nos" in arc welding.

First, he clamped the ground connection to the brake disc. This meant that the welding current passed from the disc to the axle through any convenient point of metal-to-metal contact, including the wheel roller bearings. One important rule when arc-welding on motors, pumps, tractors, or other assemblies with bearings is that you never set up a welding circuit so the current passes through bearings, commutators, bushings, or other highly finished parts which run against each other. The electrical contact will generate tiny arcs which leave rough spots and ruin the surfaces eventually.

Secondly, in the car welding job above, he neglected to cover the brake discs and hydraulic hoses before welding. No matter how skilled you are, little molten globules of weld spatter can and do fly and affix themselves to surfaces where they don't belong, or, as with the rubber hydraulic hoses, cause little localized burns which could weaken the part or cause an instant fire.

All of this relates to three important things to consider before you start welding, especially on an assembled machine or finished part:

- Take time for a careful look for anything which might be harmed by spreading heat or direct, radiated heat from the arc. This includes

Carbon-arc torch offers accurately directed heat for bending. Here, a wrench handle slipped over rod provides leverage. No real force is needed. When metal reaches right temperature it will bend almost from the weight of your hand.

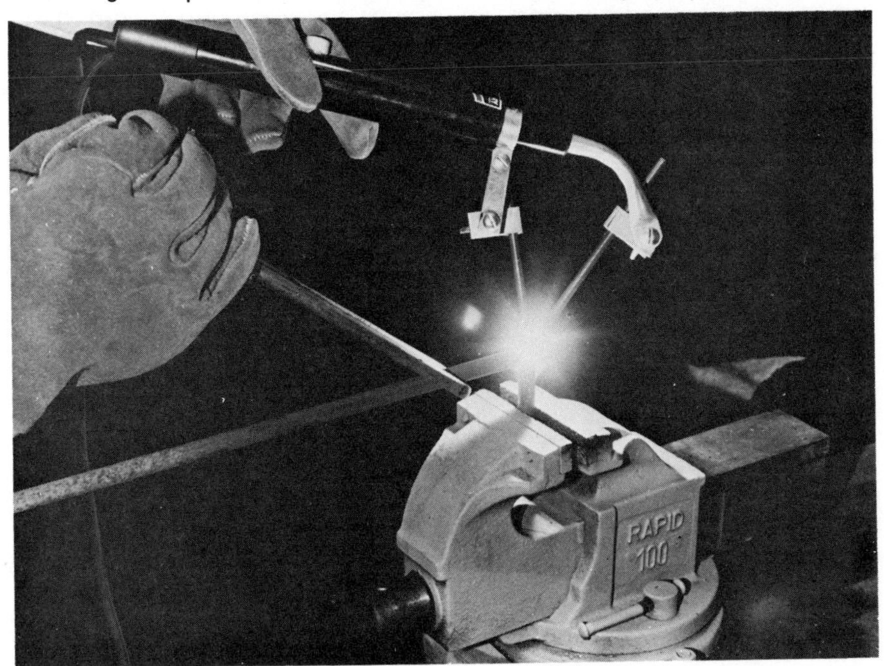

highly finished and heat-treated or case-hardened parts, oil seals, O-rings, greased assemblies, bearings, bushings, slip rings, gaskets, hose lines, wiring, electrical parts and black boxes, and, of course, gas tanks, lines, hoses, filters, carburetors, and just plain grease and drippings.

- Take time to look carefully for anything which might be damaged by arc spatter. There's no need to list such surfaces, but remember, it's easier not to spatter them than to try to polish them back to their original state.
- Take time to investigate the contents, past and present, of any container, or closed-chamber component. Never take a chance on welding a drum or other container that might have at one time held gasoline, lacquer thinner, alcohol, or a similar material. No matter how attractive using such containers may seem, enough residual material can linger for years to blow up.

Water can be equally dangerous, since the arc can turn it instantly into superheated steam. Most users of arc equipment are aware that such a heat source can be used to thaw frozen pipes. Unless you are thoroughly familiar with this technique, don't try it. For one thing your light welder may be ruined by the overload, but even worse, you can be severely injured by a steam explosion.

Small amounts of water may be where you least expect them. A friend had a 4″ piece of pipe about 4′ long with many pounds of lead in the capped end. He used it to drive fence posts and stored it beside his barn. Later, he wanted to be able to hoist this device with a rope to drive a well point and set out to weld an eye on the capped end. The resulting explosion blew the lead chunk out of the pipe; the chunk tore out the mullion of his window and went through the wall of his neighbor's house. All that

Don't make the working parts of an engine, motor, or other assembly part of your welding circuit. Example shown here would pass current through motor bearings, causing small but very damaging arcs between steel balls and races.

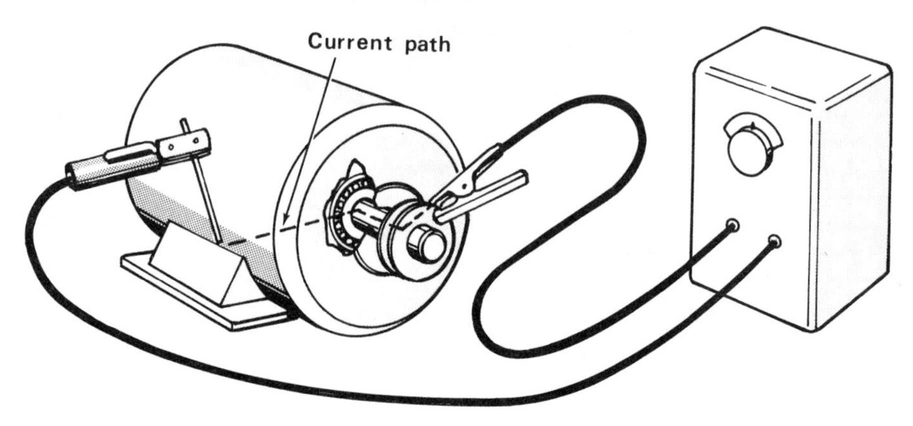

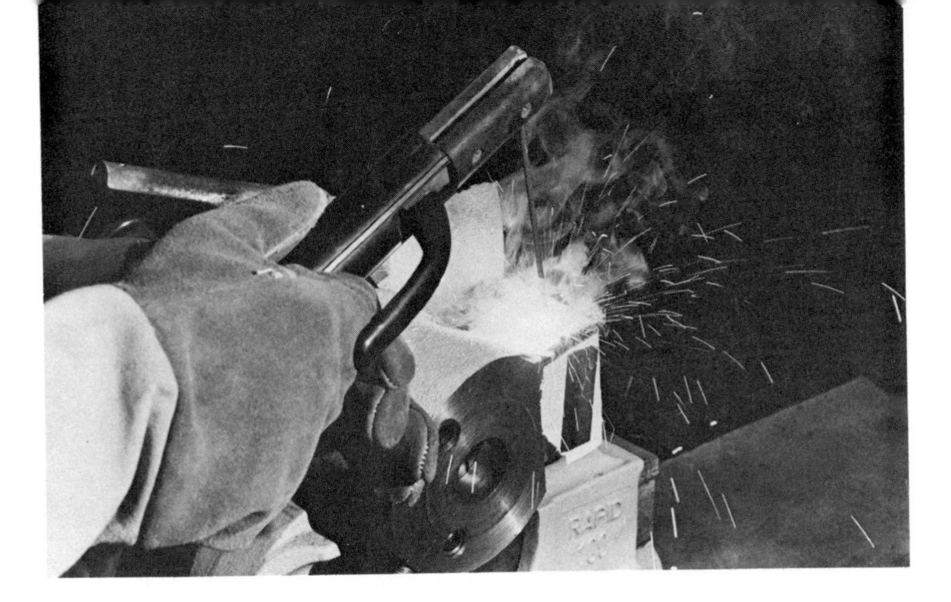

This machine-shop milling vise had some critical, polished, precision surfaces but needed welding. Asbestos paper was held in place with small magnets to expose only a very limited area for welding. If heat, rather than spatter, had been a concern, asbestos would have been soaked with water.

had happened was that a small amount of water trapped behind the lead had been converted to instant steam.

Heat shields. Many small jobs, even bench-top welding jobs, absolutely require some form of heat shielding and spatter shielding. For any number of household repairs, including the previously mentioned locking lugs on a purse, kitchen appliances, and other objects which incorporate soft solder, silver solders, or adhesives, you must have the heat confined very closely to the work area. It is disconcerting to finish a weld and find that everything else has melted apart. The quickest and easiest protection is a soaking-wet rag folded into several layers and wrapped or draped over the area. The water, because of its unique ability not to exceed 212° until it is entirely turned to steam, acts as a heat sink and picks up and absorbs heat while you weld. Arc heat, however, is so intense that it doesn't take very long to dry out a rag and start it burning.

A better material is asbestos paper or rope. Again, soak this material well and wrap or knead it into position. Since it doesn't burn you can use scraps over and over. For some structures you may want to pack asbestos furnace cement around them. This is a pastelike material. You'll also find special materials in welding shops which make high claims for their ability to absorb heat and protect work. They are probably good for some jobs, but in my experience some of the claims are exaggerated.

It is annoying to be well started on a weld and have the protective shield lift or curl away. One way to keep such shields in place is with wire, another is with spring-clip clothespins, and still another is with small magnets.

9 Difficult and Unusual Welds

You've probably noticed that I've occasionally mentioned that some welds are more difficult than others because of either their position, the metal alloy, or the metal thickness. And if you've tried some of the practice work such as welding tubular joints you've undoubtedly had difficulties making the metal deposit where you want it. The most common of the so-called difficult welds are vertical welds and overhead welds.

Attend an industrial welding class and you'll undoubtedly be trained at length in vertical and overhead welds. These are often called "position welds." That's because the welding of heavy industrial weldments presents such positions routinely. But in the home shop you may never find yourself really needing to do true overhead welding, and vertical welds will be few. This is because your project size will permit positioning your work more conveniently. If you're planning to weld ⅜" and ½" plate in assemblies weighing hundreds of pounds, this book is not for you.

You will find occasions when putting together or repairing such items as trailers, hitches, and playground equipment that it helps to be able to work from the bottom or the side. Here's how to go about it, starting first with vertical welding.

VERTICAL WELDS. By now you're well acquainted with the fact that the molten puddle flows like water. This is not a serious problem when your work is more or less flat, but inclined or vertical surfaces cause the metal to want to flow downhill. Your challenge is to learn to control or even take advantage of this flow.

There are four basic considerations in vertical welding, and for the most part they also apply to overhead welding, too:

- Electrode selection
- Polarity, with DC equipment
- Direction and technique of electrode motion
- Metal thickness

We've discussed electrodes elsewhere. If you have a choice of DC, try using E-6011 with reverse polarity. This is a fast-freeze electrode, and the reverse polarity puts most of the heat into the rod and less into the work. This helps toward quick solidification so the metal has less chance to run. If you have only AC to work with your best choice is E-6013. Remember, the more metal there is in the puddle the greater the tendency to run. Therefore, you will do better to use the smallest rod that's practical and run several beads if necessary.

Welding "up." The direction and technique of electrode movement involves the thickness of the metal. On heavy work such as ¼″ and thicker plates it's considered good practice to start at the bottom of the joint and weld up. This gives deeper penetration. In welding up you start at the bottom and deposit small amounts of metal. The idea is to let each deposit cool slightly before adding to it. The technique used is called "whipping," because you watch the puddle form and then whip the tip of the rod upward slightly to start the metal cooling before it has a chance to run. Follow-up welds after the first bead are then made by weaving the rod back and forth slightly to broaden the first bead. Again, the trick is to let the deposit cool slightly between weaves. If this sounds a trifle difficult, it is.

Welding "down." I recommend a somewhat easier method that will probably be all you ever need in the home shop. For a beginning, place two pieces of ⅛″ or 3/32″ steel to form a butt joint or a fillet joint. Tack them if you wish, and then position them on the bench to slope upward at about

Pro welders often use a "whipping" motion when vertical-welding heavier metal and working "up." Momentary upward whip of rod tip permits deposited metal to cool before arc is dropped back down to deposit more metal.

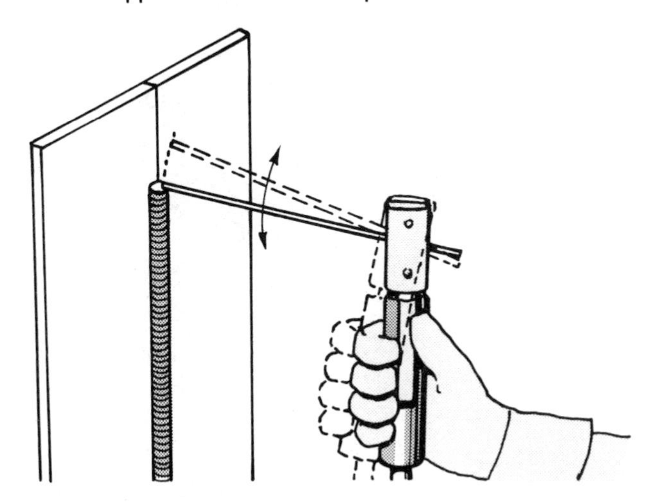

a 45° angle. The angle is a little easier than a full vertical for your first practice. Your basic technique will be to produce a stringer bead, in this case welding downhill.

Again, your best choice is an E-6011 rod, reverse polarity, if you have DC. The electrode movement for this type of bead is called "drag welding" because you're really dragging the molten puddle along behind the rod tip as you move the latter downhill. Moreover, you're going to allow the edge of the rod coating to actually contact the metal surface ahead of the puddle. This means holding the electrode at a much flatter angle to the metal. About 60° to 80° is suggested by some; others recommend an even flatter angle of about 30°, and I've found that this is easier for me. The object of touching the rod coating to the metal is to form a little blockade to prevent the molten puddle and slag from running down ahead of the rod tip. The force of the arc and the gases coming from the rod tip tend to push the puddle and slag back uphill.

You'll discover that as soon as you've started the arc and established a puddle you'll have to move the rod tip quite fast to keep ahead of the molten slag which follows right behind the puddle. If slag runs ahead of the electrode the weld gets messy.

With lighter metal where heavy bead is not necessary, welding "down" with drag technique usually works better in home shop. Practice at 45° running stringer beads and position work at successively steeper angles as you gain skill.

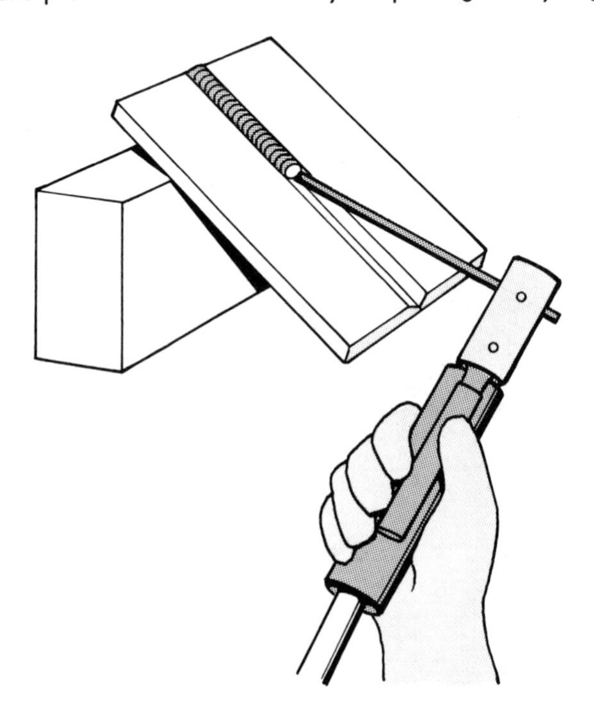

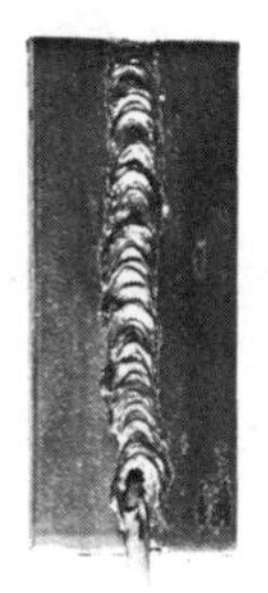

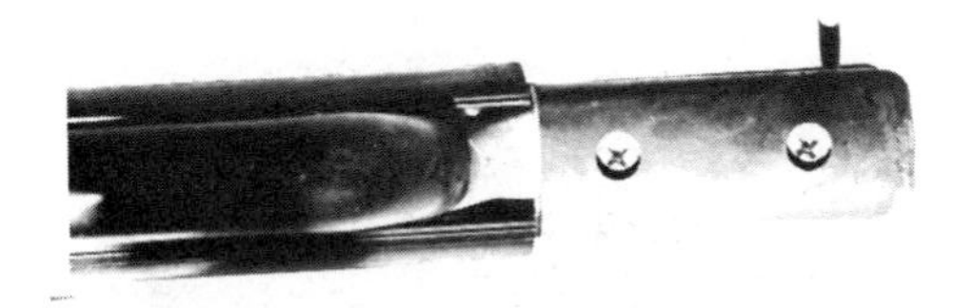

Vertical drag weld was made "down" with electrode at a very shallow angle and rod coating actually dragging on surface of metal. Slag has been removed and holder positioned to show result and how it was done. Method is good for light metal.

Practice such fast runs to produce stringer beads until you get the feel of the proper rate of electrode and puddle movement. Then gradually increase the angle of the workpiece to vertical until you are finally able to run a true vertical weld with confidence.

OVERHEAD WELDING. Sooner or later you'll find that you must get under something and deposit a weld bead above your head. It should be obvious that in addition to the difficulty of making molten metal stick in the joint rather than fall out you must take more heed than usual for personal safety. Sparks and globules of metal have an uncanny way of finding their way into the open cuffs of your gloves, down your neck, and into your hair.

Watch a pro weldor handling an overhead weld and you'll probably see him position the rod in the holder at an angle which permits him to keep his gloves turned so the cuffs are down. He'll probably wear a hat to keep sparks out of his hair, and he'll button up his shirt collar. A leather jacket is recommended. He may also place the electrode cable over his shoulder to reduce the drag.

To practice overhead welding, clamp a practice piece of ⅛" steel to the top of a stepladder or other handy support and start by running beads much as you did when first starting downhand welding. Once more, reverse polarity and E-6011 rod are recommended if possible. Use a ⅛" or lighter rod and try to run a straight bead without weaving or whipping. The trick is to move steadily and fairly fast to prevent the metal from sagging. *Caution:* If you have cleared only a limited area of inflammable material away for your practice, take pains to clear at least twice that area before starting overhead practice. Be assured that there will be many more sparks and bits of hot metal bouncing around the floor.

When you've mastered the basic technique of running a bead overhead, try clamping two plates with a slight gap between them so you can try a butt weld. To get the needed penetration you'll probably have to resort to a moderate whipping action to get the puddle down into the joint and

Practice overhead welding in a clear area or outdoors. Expect plenty of molten metal and sparks on the floor. Button up tight and pack in some extra protection at the collar to prevent burns. Here, it would have been better to bring welding cable over weldor's shoulder.

then let it cool slightly. This technique may not work as well if you continue practicing and go on to a tee weld overhead. Here, since you need less filler material to make the joint, a straight, fairly light bead will probably work better. You may also find that a little higher amperage will help on such fillet welds.

CRITICAL METALS. So far most of our discussion has centered on welding mild steel. In the past, mild steel was widely used in farm machinery and automotive construction. Today, in the interest of high strength and lighter weight, more and more steel with higher carbon content is being used. And some high-alloy steels are becoming common where you might not expect them. Ordinary, general-purpose electrodes such as E-6011 and E-6013 may produce visually acceptable but actually unsound welds in these materials. Typically, the defects will be subsurface cracks not readily spotted by the home-shop weldor.

Ideally, you should first identify the steel—its alloy characteristics, carbon content, and hardness, and whether it has been heat-treated. In practice this is seldom possible for the home-shop weldor making a once-in-a-lifetime repair on a given piece. On the other hand, the professional weldor's livelihood and reputation depend on his handling of such jobs. He has access to trade journals, bulletins, and the grapevine of information that develops in any specialty trade. For this reason I strongly recommend taking critical welds on expensive machinery or on structures where failure could be life-threatening to a pro.

In other cases, an E-7018 electrode, called a low-hydrogen type, will usually produce satisfactory results. These rods generally require reverse polarity and fairly high amperage. This rules out the straight AC welder and many light-duty welders. If you want to try this type of rod, specify that you want it for "repair welding."

I repeat, if you suspect that your job involves high-strength steel and if failure could have serious consequences, do not try do-it-yourself repair welding!

HARDFACING. The life and performance of many tools used to dig, plow, or move soil can be extended and renewed by applying super-hard alloys to the workface by arc welding. There are many hardfacing products on the market, and you should explain to your supplier what you want it for and what the amperage characteristics of your welding equipment is when you buy. The basics are about the same with respect to procedure.

The surface should be thoroughly cleaned by grinding. Place and secure the workpiece so it is flat and in a position for downhand, horizontal welding. If a part is badly worn it may be necessary to build up the worn area with mild steel rod, or even to weld in a patch, before hardfacing. Normally, plan to use ⅛" electrode on material up to about ¼" thick and hold amperage to no more than needed for good flow. You don't want the hardfacing material to penetrate the workpiece too deeply and destroy the characteristics of the base metal. Try to produce a relatively thin layer of hardface material—no more than 1⁄16″ to 1⁄8″ thick—and weave the rod to develop a wide bead 3⁄4″ to 1″ wide. Side-by-side beads are used to cover broad areas. Since the hardfacing is too hard for subsequent machining, you should try to avoid excess deposits. If you have some lumps they can probably be ground away, but not readily. Most suppliers recommend hardfacing with straight polarity and suggest a "spray-on" technique with the arc gap held longer than usual.

CAST IRON. Welding cast iron can be very successful and it can also be very disappointing and frustrating. Much depends on the nature of the cast iron involved. The main problem is that cast iron is very brittle and does not accept the expansion and contraction of welding as does the more elastic mild steel you've been practicing with. It is not uncommon to be standing back admiring your weld in a piece of cast iron only to hear it pop and crack as it cools. In some instances it may be necessary to heat the entire casting to an even dull-red heat with either a gas torch or an oven before welding. Afterward the casting should be buried in dry sand or dry slaked lime and allowed to cool slowly over many hours.

In most cases, however, cast iron may be welded with machinable nickel rod. This rod leaves a soft bead which may be filed, ground, or machined to a final surface. One maker of such nickel electrodes advises that you "weld a little and cool a lot." What he means is that you should not attempt to run a bead the full length of any long crack in one or even two passes. This will become clearer as you read on.

Before welding, the crack should be cleaned with solvent, by brushing, or whatever is needed to remove paint, oil, and other contaminants. A chisel or grinder should be used to open up the crack for its full length and a little beyond. Cracks have a way of extending past their visible

Typical cylinder-head repair would be a freezing crack with no totally loose pieces. Drill at extreme ends to interrupt crack. Vee-out crack to about half of metal thickness. Run short beads and peen each gently as it cools. Nickel rod permits easy dressing of bead later.

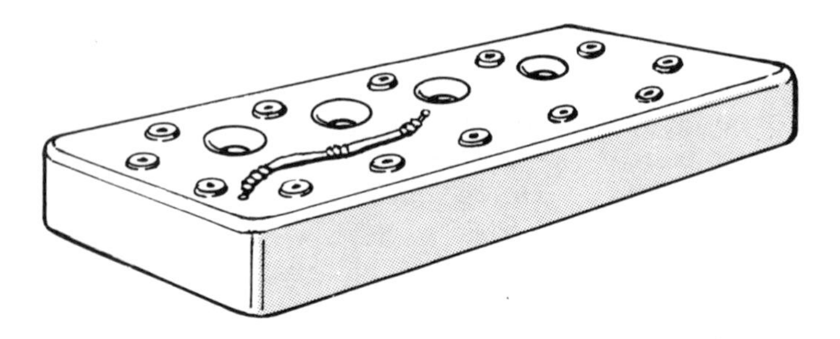

This sector gear was broken off at right-angle portion. Repair was feasible only if relationship of central shaft hole to bolt holes could be maintained together with alignment. Shaft support piece was turned to fit, spacer collar cut to length, and holes drilled in scrap-metal radius arm at proper distance. Two heavy metal blocks, one under gear teeth, other at outer end, completed set-up. Initial tacks were made by carbon-arc brazing at outer corners. Final weld was made with nickel rod, both sides.

ends. For this reason, if at all possible, many weldors will drill a small hole, 1/8" or 3/16", at each end of the crack to inhibit the cracking from starting again.

If the crack is in a structure such as a water jacket, pump housing, or cylinder head, the most important goal is that it hold water. Thus, you can probably do a better and easier job if you groove out the crack to only half the thickness of the metal. Opening the crack all the way through introduces the danger of the weld puddle wanting to drop inside the casting.

Support the iron. If the crack is a full fracture so you have two or more pieces to cope with, you must somehow arrange to hold these pieces in good match and alignment while welding. Before starting, give some thought to how the part functions and what dimensions are critical. For example, does it have bearing or bolt holes in each of the broken parts which must be spaced when finished to align with mating holes? Is alignment, as of two bearing or bushing bores, critical? Will a buildup of weld bead interfere with assembly or operation?

The ragged edges of the break are your best guides for alignment, so, although you should grind back the broken edges to form a vee for welding, you should not remove so much metal that there's nothing left to match. Try to leave at least 1/16" of the original surfaces for matching.

Usually, broken castings are irregular in shape and hard to clamp in alignment. One method is to fill a pan with dry sand and bed the parts in it for tacking. Often you'll want to tack and weld small beads on first one side and then the other to keep distortion effects balanced. The sand support works well here, since you must allow your welds to cool thoroughly anyway and there's no problem in rearranging the sand to support the other side. In some repairs, just supporting the part is not enough. You must physically hold the spacings of holes or other features to the exact original dimensions. About the only practical way to do this is to make up a supporting structure or jig which includes the holes involved and which can be bolted or pinned in place so nothing can shift while welding. Such methods are time-consuming but may be the only solution in, for example, restoration work on old equipment with parts no longer available.

Cooling is vital. I repeat, it's very important when repairing cast iron to deposit short welds and let them cool thoroughly before welding again. In most cases of fairly long cracks you must run a short bead at one end, another short bead at the opposite end, and another in the middle. Allow each bead to cool until it is comfortable to hold a bare finger on it. As soon as the weld is made and before it cools significantly, use a light cross-peen hammer to peen the weld metal lightly to prevent the buildup of stresses. Do not strike too hard and do not strike the edges of the iron.

Make your next weld only after peening and cooling. This procedure is slow, of course, but any attempt to speed up repairs on cast iron will almost always result in more cracking. Never use an air jet, water, or other means to speed cooling. Finally, before setting out to arc-weld a casting, give consideration to the possibility of braze-welding it with a carbon-arc torch. Brazing is usually easier, and since you're working below the iron's melting temperature you're less likely to destroy thin sections of metal. Refer to the chapters on the carbon-arc torch and brazing.

10 Special Arc Welding Tricks

So far, with the exception of cast-iron repair and hardfacing, we've stayed very close to the basic process of welding two pieces of mild steel together with a bead or fillet. But there are other tricks possible with the electric arc which are extremely useful. These include cutting, piercing, gouging, removing bolts, rivet welding, thin-metal welding, spot brazing, brazing, and aluminum welding. The last three are done with the carbon-arc torch and will be discussed in Chapter 11.

CUTTING WITH THE ARC. By now you've burned through enough times to realize that the electric arc can penetrate metal easily. Cutting is simply a matter of controlling and directing the burning-through process. With practice you can make cuts almost, although not quite, as smoothly as with the oxyacetylene torch (see Chapter 18). Although ordinary mild steel rod such as E-6011 can be used for cutting, a special cutting rod is recommended. And for really good cutting action, high amperage is required. In fact, some welders have a special cutting tap. One advantage to arc cutting is that it works just as well on extremely hard material such as spring steel as it does on mild steel. Also, it can cut heavy plate ¼″ to ½″ or more thick, which would be quite a challenge to a hacksaw.

Splatter precautions. Before starting any cutting operation you should take extra precautions against personal burns and starting a fire. Remember, you are going to produce considerable molten metal which will fall out of the cut and spatter broadly. If at all possible a bucket or pan of dry sand should be placed under the cutting area to catch this waste metal. Be sure that your trousers are of full length, preferably of wool, and that you have no openings which would allow a globule of molten metal to enter your shoes.

Even heavy steel is readily cut with arc. High amperage is needed for thicknesses like this. It helps to use special cutting rod. Action of welder is almost the same as using a hacksaw.

Cutting technique. The actual cut is easily made. Using full amperage and a ⅛″ or 3/32″ rod, start by striking an arc on an outer corner of the workpiece. Straight polarity is advised if you have DC available. On stock ¼″ and thinner it's customary to start on a top-surface corner. On heavier stock, start on a bottom-surface corner. Once the puddle is well started you simply push along the cut line with an up and down action like short hacksaw strokes.

Piercing holes. The same burn-through method can produce remarkably neat holes. This can be very handy if the metal is too hard or would require a very large drill. To pierce a hole you strike an arc and hold it until the puddle is well established. Next, push the rod right on down through the metal. If the rod sticks you pushed too soon. To enlarge the hole, draw back the electrode and hold a long arc to work around the edges until you

Hard, thick spring stock such as this would be almost impossible to drill, yet it's pierced in seconds with arc. It takes practice, however, to finish up with just the right-size hole.

have the diameter needed. The same basic approach can be used to melt off a frozen nut or bolt head.

Gouging. Very often it is necessary to remove old weld metal or to open up a vee for repairing a crack, but chiseling or grinding is tedious or impractical. Or you may need to bevel the edges of heavy plate for welding. The arc may be used almost like a gouge chisel in wood to open up a clean groove or trim off an edge. The trick is to use straight polarity, if DC is available, a cutting rod, or even E-6011, and hold the rod at a very shallow angle, almost flat to the work surface. After starting the initial puddle, feed the rod across and into the desired groove and allow the force of the arc to blow the molten metal out ahead. If you try to go too deep on the first pass, you may pierce the work, so hold a shallow angle and repeat as needed for added depth.

RIVET WELDING. Construction or repair with thin sheet metal has always been a special challenge to the arc welder. Burn-through and distortion can be very frustrating. In many cases a perfectly complete,

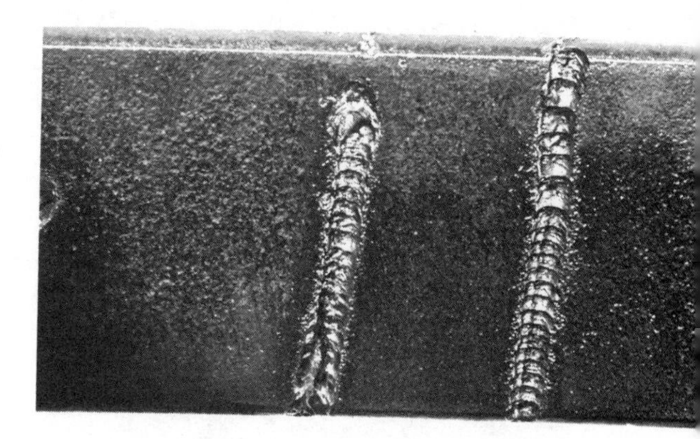

Another example of arc gouging shows how you can control depth and direction quite accurately. Trick is to keep cutting rod almost flat on work metal and let hot gases force out slag.

full-length seam is not required; in fact, a great many automotive seams are only spot-welded. Lacking such a spot welder it is perfectly possible to stitch two pieces of metal together by rivet welding. Since such welds are made very quickly and in a limited area, there is practically no warpage or distortion.

Rivet welds are made by pushing rod through puddle quickly and withdrawing slowly. Finished weld is seen at right. Posed picture, top, shows rod action. Action photo shows that rod has pierced the two sheets of metal and most of heat is now underneath. It will deposit "rivet" when rod is withdrawn.

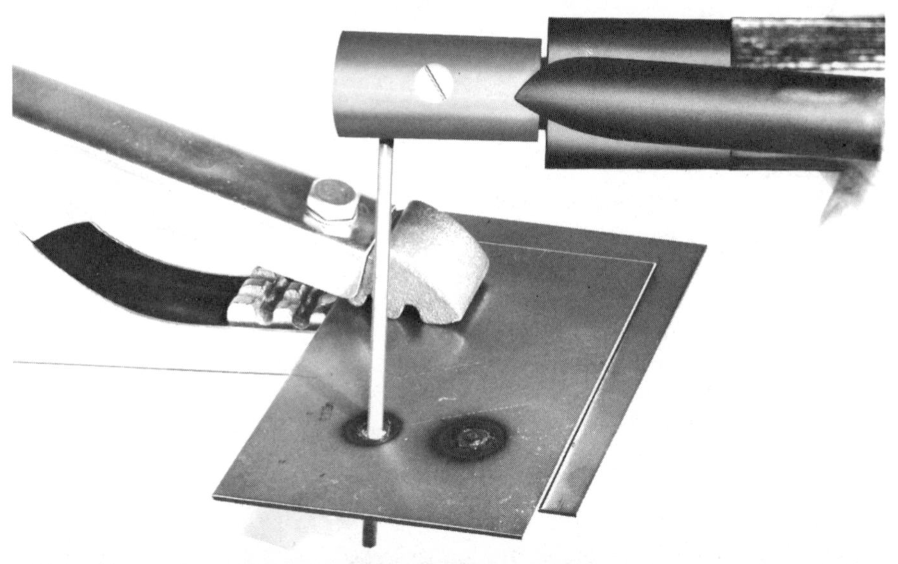

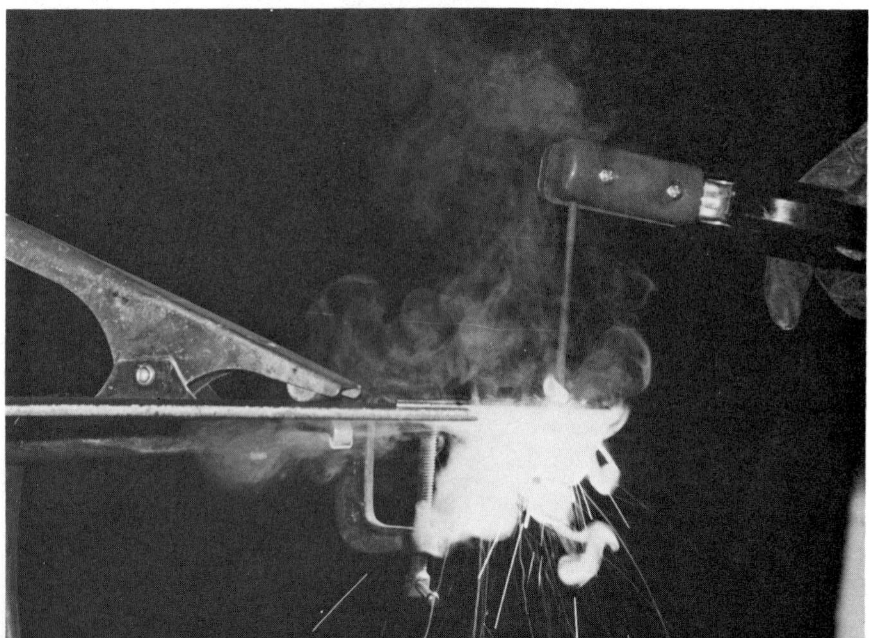

Seams in $\frac{1}{16}$" sheet steel such as this are practically impossible with some welders. Recently, however, improved internal circuits have made it easy to strike a soft, low-amperage arc and not burn through.

To make a rivet weld, first overlap and clamp the pieces together and in firm contact at the weld area. You don't want a gap or air space between them. Use E-7014 iron-powder rod and about 50–100 amps, preferably reverse-polarity DC. Strike an arc and as soon as the metal starts to puddle push the rod right through the soft spot. Withdraw the rod slowly and you'll deposit a button of metal in the hole that very much resembles a rivet. The entire process takes only a second or two. Generally, $\frac{5}{64}$" or $\frac{3}{32}$" rod is adequate. A series of such rivet welds can button together thin sheet metal very neatly.

WELDING THIN SHEET METAL. Sometimes it's absolutely necessary to weld thin sheet metal all along an edge, or to attach a thin sheet to a heavier section. But not all transformer welders are equal in their ability to handle conventional bead welds on thin steel such as automobile bodies. Thus you may find it almost impossible to strike and hold an arc and still not burn through with some low-cost, low-amperage AC welders. Part of this is simply that the low amperage required makes starting the arc very difficult, and any boost in amperage to the point where you can hold an arc is too much and burns through as soon as you start to weld. This relates to the internal design and open-circuit voltage of the welder and there isn't much you can do about it.

Other welders are better-behaved and with 35 amps, E-7014 rod, and reverse polarity it is very easy to strike a soft burning arc and weld thin metal neatly. Some weldors will disagree with the use of reverse polarity, since it puts most of the heat into the rod. My experience is that it works just fine, but if straight polarity works better for you, use it.

11 Using the Carbon-Arc Torch

As mentioned earlier, the carbon-arc torch is the best heat in the house. It's quick, cheap, and hotter than any other source. Even better, the flame has no blast or pressure, so light pieces are not blown around. You'll find it useful for everything from soft soldering to silver soldering,. brazing, aluminum welding, heating and bending, loosening rusted nuts, and tempering tools. Torch carbons are hard, brittle, and break easily but are cheap enough to be considered expendable. Used within the recommended amperage ranges they last quite a while, at least for several jobs in most cases. When buying you must select either AC or DC carbons, since they are made differently. The usual AC carbon has a copper coating which helps to conduct current down to the tip.

Recommended Carbon Sizes and Currents

1/4"	30–40 amperes
5/16"	40–65 "
3/8"	65–90 "

Self-protection. Because the carbon-arc torch is so handy to pick up and flip on for so many jobs, some users tend to get casual about protecting themselves from its intense rays. There's a tendency to wear only gas welding goggles and work with bare hands and arms. Never get into such dangerous habits! The ultraviolet radiations can cause severe burns and skin cancer, and even destroy eye tissue. Use a regular arc welding helmet with at least a #11 or #12 lens. A #14 is not too dark if you are using heavier carbons and higher amperage. Wear a long-sleeved shirt or jacket and leather or other heavy work gloves. Don't adjust or strike the arc

Changing angle between carbons alters shape of arc flame from wide, top, for brazing fillet, to narrow, bottom, for executing a crisp, sharp bend in strip steel. Take advantage of the fan shape to work into corners. Job on top will go best if torch is turned 90° to put fan of flame parallel to fillet.

without your helmet down, and don't place the torch on a metal surface or bench while the current is on. It's a good idea to make up a little hook or bracket on which to hang the torch while you're adjusting the work. Remember, the carbons will remain very hot for some time.

The arc flame. The carbon-arc flame is a soft, fan-shaped pattern with an inner and an outer heat zone. The inner zone is close to the carbons and provides a temperature of 9000° F. or more. The outer zone is less intense and tends to heat a broad area quickly. To some extent the flame shape can be broadened or narrowed by changing the angle of the two carbons. You can also vary the flame by altering the distance between the carbon tips. In general, the best flame has a soft, purring, rather quiet sound. Try adjusting the thumb slide of the torch to learn to recognize the sound of a properly adjusted arc.

Since the arc flame is flatter than the flame of a gas-fueled torch it is important to orient the flame to the joint. If you're working down into a corner or fillet, hold the torch to aim the flame parallel to, rather than crossways to, the corner.

GETTING STARTED WITH CARBON-ARC. As delivered, carbons are usually bluntly tapered at the contact end. If you're using a carbon that's been broken off you should taper it this way with a bench grinder or a file. The carbon torch, except for the little chopstick torches, usually has one carbon holder side fixed and the other one movable. By sliding the thumb button on the handle you can bring the tips of the carbons into contact to start the arc and slide (or rotate on some torches) them apart to adjust the flame. This provides a handy way to start and stop the torch instantly.

Carbon adjustment. Set the carbons in the clamp-type holders so they extend at least 2″ or 3″ below the clamps. Never do this with the power turned on. The carbons must be set so they will contact for starting, and you'll want to go a little tighter than this so there will still be contact after you've burned away some of the carbons. Sometimes the torch parts get a trifle loose and the carbons misalign so they brush past each other or don't contact at all. Take time to see what needs adjustment and snugging up so you get a nice square taper-to-taper contact. Misaligned carbons distort the flame and make your work more difficult.

Carbon torch hookups. Most carbon torches have two leads. The most convenient arrangement is to have terminal connections which plug directly into your welder terminal taps just like your ground clamp and electrode holder. If you acquire a torch with terminals which don't fit your welder I strongly recommend buying two terminal fittings which do fit to replace the originals. Some weldors don't bother with this and simply clamp the misfit terminals in the ground clamp and electrode holder. This works for occasional use but it puts a lot of unnecessary wire in the system and sooner or later the ground clamp and electrode holder get kicked together and start a little private arc of their own. By Murphy's law such a short circuit will occur right under your trouser cuff or off in a remote corner where it can start a fire or burn the insulation off your cables. Some welders have a fixed ground-cable connection and you'll have to hook to the ground clamp, but if you can plug the other terminal from the torch into the welder tap you've at least eliminated the fire and short-circuit hazard.

Note that the typical torch set-up I've described passes the current through the torch and the circuit is completed between the two carbons. Thus the arc flame may be struck and held independently of the work much like a gas torch. This permits you to adjust the flame before starting your work. It also provides enough light so you can position the flame and filler rod before starting. But all carbon-arc work is not done this way. In some cases you'll want to clip one side of the circuit to the work and the other carbon to the welder exactly as you do for metallic arc welding. We'll describe such techniques later.

Heating with carbon-arc. To understand and appreciate the capabilities of your carbon-arc torch try a few practice runs using it simply as a heat source. Clamp a scrap piece of ½″ rod, strap iron, or whatever is handy in the vise and mark a point an inch or so above the vise jaws. Start with ¼″ or 5⁄16″ carbons. Forty or 50 amps will probably be enough. After being certain that you've protected yourself against arc burns, separate the carbons, pull down your helmet, and flip on the welder. Slide the thumb control to bring the carbons into contact. The arc will strike instantly and have a crackling or sizzling sound. Separate the carbons to produce a soft, purring flame.

Now bring the flame into the position where you've marked and bring the inner flame in close, but not touching the mark. Expect the heat to bring the metal to a red temperature quite quickly. Experiment by bringing the inner flame right in close to the metal. You'll see the metal start to appear wet, maybe bubble and blister, or even melt away, if you hold this position too long. What you've learned is that in cases of fairly heavy metal the intense heat may be more than you want on one side. It can easily damage the metal before the other side or the interior warms up. The trick is to move the torch from one side to another to bring the metal to an even heat throughout. Suppose you wish to produce a sharp bend at the point you've marked. The easiest way to do this is to slip a piece of pipe over the workpiece or grip it with a wrench so you can apply a modest bending pressure while you are heating. This avoids the need to guess as to when you've reached bending temperature, since in most cases you'll tend to overheat when guessing. With the pipe or wrench you'll sense immediately when the metal starts to soften, and in another few seconds it will yield like taffy. With a little practice you'll be able to produce sharp bends, gentle curves, twists, or whatever shape you wish. If you wish to heat the metal so you can forge or hammer it, be prepared to switch off the torch, lay it down where it won't start a fire, and tip up your helmet quickly. This type of work requires some sort of solid anvil and a machinist's or blacksmith's hammer plus heavy pliers or tongs to hold the metal. Here you will probably want to heat the metal over a broader area than for a sharp bend, so you must observe your torch movement, flame position, and metal color as heating progresses.

Hardening and tempering. Another thing you'll find easy to do with carbon-arc is hardening and tempering such tools as cold chisels and punches. Since this is a topic in itself involving the identification of metals and choice of cooling materials and procedures, the following brief description is too limited to do more than give you an idea of the technique.

Assuming, for example, that you have a cold chisel and wish to sharpen and harden it, the first step would be to clean up the cutting edge to approximate shape with a grinder. Remember, you only want to harden a fairly short portion of the cutting end. Above this end you want moderate hardness and toughness. You do not want to harden the end you hammer, since this could cause dangerous chips to fly off when struck.

Support the chisel in a vise or otherwise so you can work the arc flame under, over, and around it evenly. Start by heating the lower two-thirds or so to a cherry-red heat with your torch. Some heat the entire chisel, but it's not necessary. Quickly plunge the chisel into a quench bath of water

Carbon-arc torch can harden and temper tool steel such as chisel shown gripped in vise. Move torch along and around work for even heating clear through. Keep pliers and quench water handy for quick action.

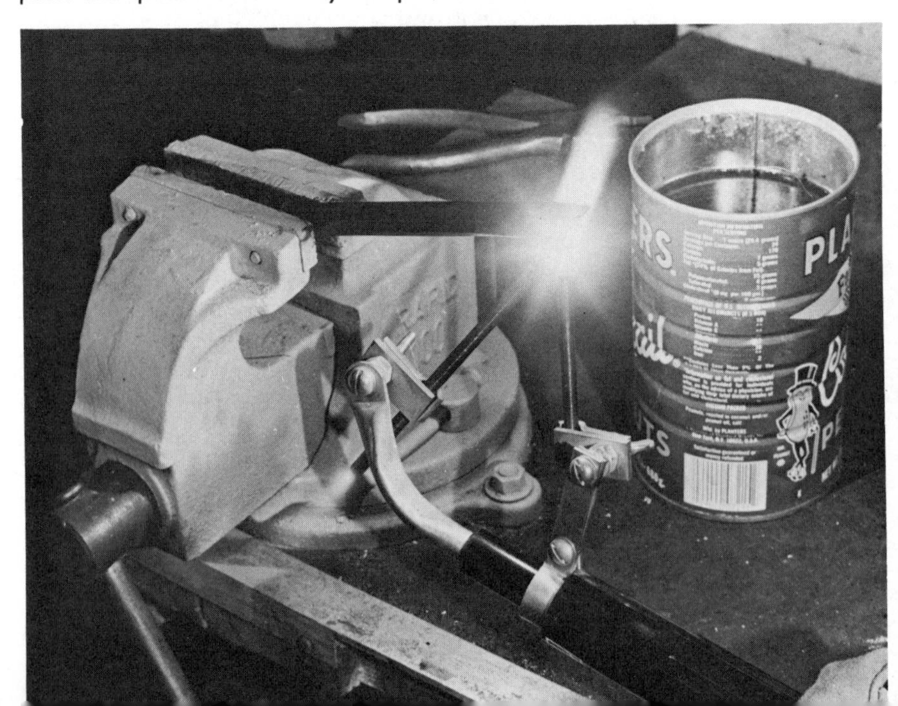

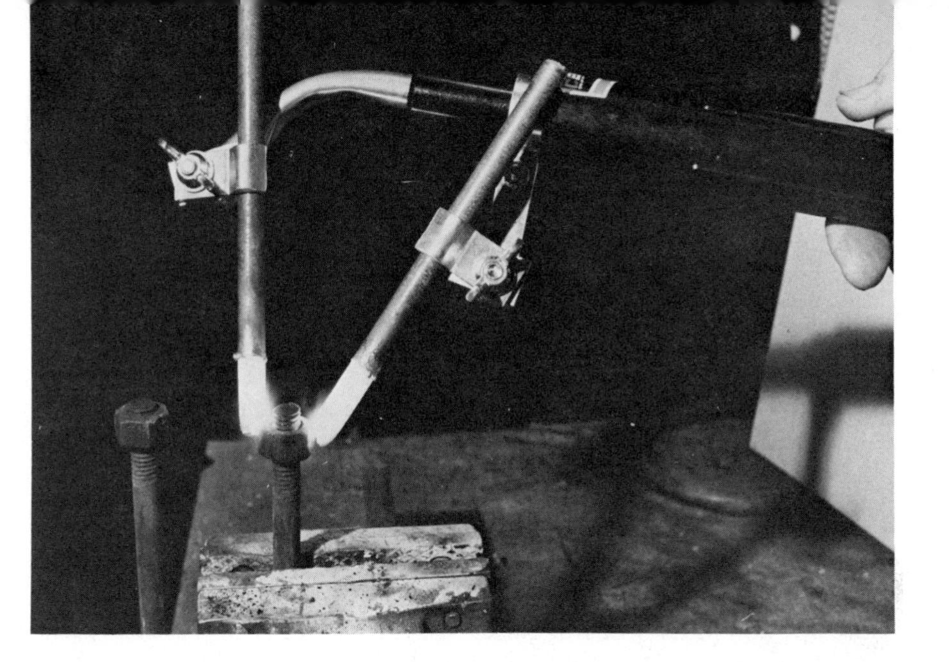

Even badly rusted nuts and pipe fittings can usually be loosened without damage by placing carbons across them. No flame is needed. Carbons will glow as shown; so will nut if you hold position long enough, but it's seldom necessary.

to a depth of about 1″. When cool, the cutting edge should now be glass-hard, and running a file over it should prove this. This is too hard. The metal must be tempered to avoid breaking in use. Tempering can be done either by heating the portion of the chisel behind the cutting edge or, after gaining experience, by working quickly after quenching the cutting edge and while the main portion is still red.

To temper, polish the cutting face of the chisel to a bright surface with abrasive paper or a power sander. You can do this quickly while the main body is still hot. Colors will start to creep from the heated portion toward the quenched end. These colors will range from deep blue and violet to an almost invisible off-white. They will move quite quickly along the chisel. Your challenge is to spot the arrival of the first light straw color, then a deeper brown, and finally a deep-purple color at the cutting edge. Immediately, but gradually, quench the cutting end of the chisel in water. When the rest of the chisel cools below dark red, immerse it in water.

Freeing rusted parts. The intense heat of the carbon-arc torch can be used, of course, to melt away stubborn nuts and bolts very quickly, but often you do not want an open flame which could be dangerous or damage surrounding parts. And you may not want to destroy a part if you have no replacement.

Here, the carbon torch may be used as a flameless heating device simply by bringing the two carbon tips into firm electrical contact with each side of the nut and turning on the power. It is helpful to file two bright spots through the rust to get good contact. The carbons will quickly start to glow and the nut will heat rapidly to a red color. Normally this degree of heat is not needed. Try heating the nut just below red and quickly applying wrench pressure. The heat and expansion usually do an amazing job of freeing otherwise immovable nuts.

BRAZING. The terms "brazing" and "braze welding" have been twisted about freely in the language of joining metals. Fundamentally, both involve heating iron or steel (sometimes other metals) to a temperature much below melting or puddling temperature and using a lower-melting-point alloy such as bronze to join them. Usually about 1000° F., or in some cases a dull-red temperature, is hot enough for such joining. This is not much different from soft soldering except that the joint is much stronger. Very often a brazed joint is preferable to a welded one. An example might be the light tubing of a cycle frame, light car-body metal, or a cast-iron gear or engine jacket. In the case of the cycle or car body it's too easy to burn through and seriously damage the metal unless you're an expert and your welder handles thin metal well. Since brazing is done below the melting temperature of the work metal, it's safer. In the case

Light tubing and conduit are more easily joined by brazing than by welding in most cases. Carbons here are at a narrow angle to hold flame to a small area so completed portion of bead cools and does not remain fluid. Gloves were omitted to permit camera operation.

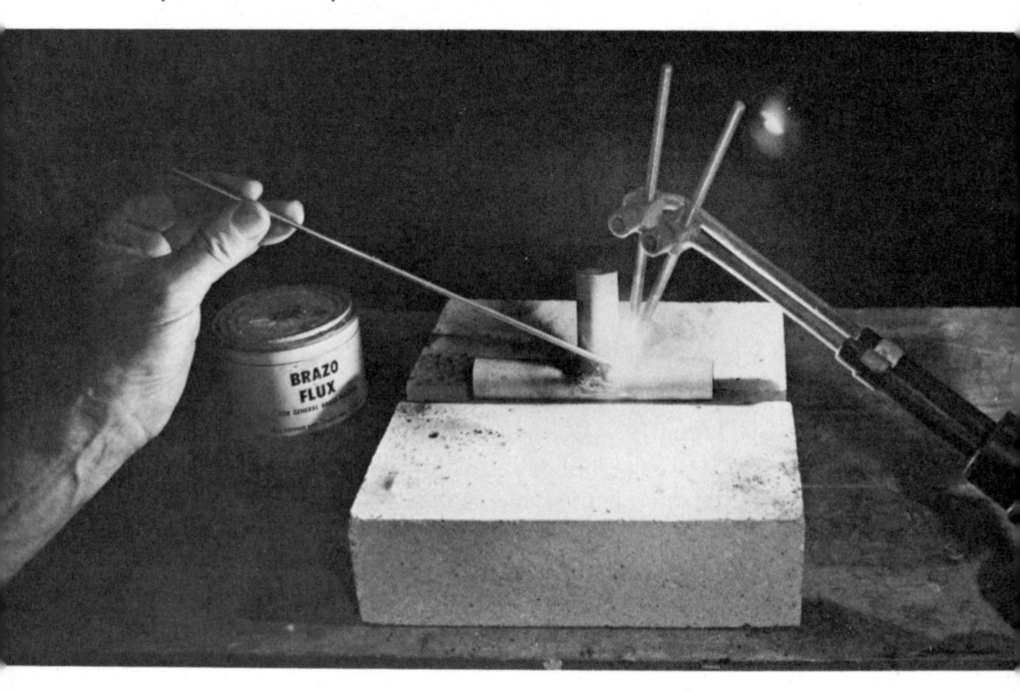

of cast iron, the lower heat and the broader flame of the carbon-arc torch used for brazing, as opposed to the high, concentrated heat required for welding, help reduce the tendency to crack. The carbon-arc torch is ideal for most such brazing jobs, including some very small ones and some rather large ones.

Brazing rod comes in nominal 3′ lengths and 1⁄16″, 3⁄32″, and ⅛″ diameter. Such rod is sold without any flux coating, and an inexpensive borax-base white powder flux is needed to clean and protect the workpiece so the metal spreads and adheres. There are other so-called "pre-fluxed" rods sold with coatings or metallurgical characteristics which make them self-fluxing for many jobs, and these are convenient to have on hand. Most such rods cannot be readily identified by any visible coating, although they tend, at least with most products, to be of a more reddish, copper color than the bright brass brazing rod for use with flux. Even so, there will be times when you need a bit more fluxing action, and a can of brazing flux should always be handy on the bench.

Practice brazing. Before starting a braze joint it is important to clean away dirt, paint, rust, and scale so the metal is bright. Use a wire brush, file, abrasive paper, grinder, sander, or other means to cut down to clean, bare metal. After doing this I ordinarily sprinkle some brazing flux around the work area. This serves to clean and protect the metal during the warmup to brazing temperature and has the advantage of melting into a liquid, glasslike state when brazing temperature is reached. Normally 40 or 50 amps and ¼″ or 5⁄16″ carbons are about right for most home brazing jobs. As always, protect yourself and your work area from ultraviolet and arc flame before starting. Apply heat in a somewhat brushlike manner to the work area unless it is necessary to really confine it because of surrounding material or structure. At the same time pass the arc flame over the tip of the brazing rod for a second or two and dip the rod into the can of flux. If the rod is hot enough you will withdraw it with a coating of brazing powder at the end.

Place the end of the rod in the outer flame area, but do not try to melt it until the flux you sprinkled on the joint earlier turns liquid. When it does, introduce the rod tip and allow the braze metal to melt and flow into the joint. The metal should flow smoothly and readily. It should also follow the heat if you move the torch along the joint. Do not overheat. Sputtering and popping usually mean that your work is too hot. It may also mean that some contamination is present. You'll find that by judicious balancing of heat and adding rod you'll be able to build, fillet, fill gaps, and even add on and reshape areas of work that are worn or broken away.

Brazing cast iron. Brazing normally works as well on cast iron as on steel, but, as with welding, you must be very careful to avoid cracking. There are several ways to do this, depending upon the shape and nature of the work part. The broad, fanlike nature of the arc flame permits you to brush

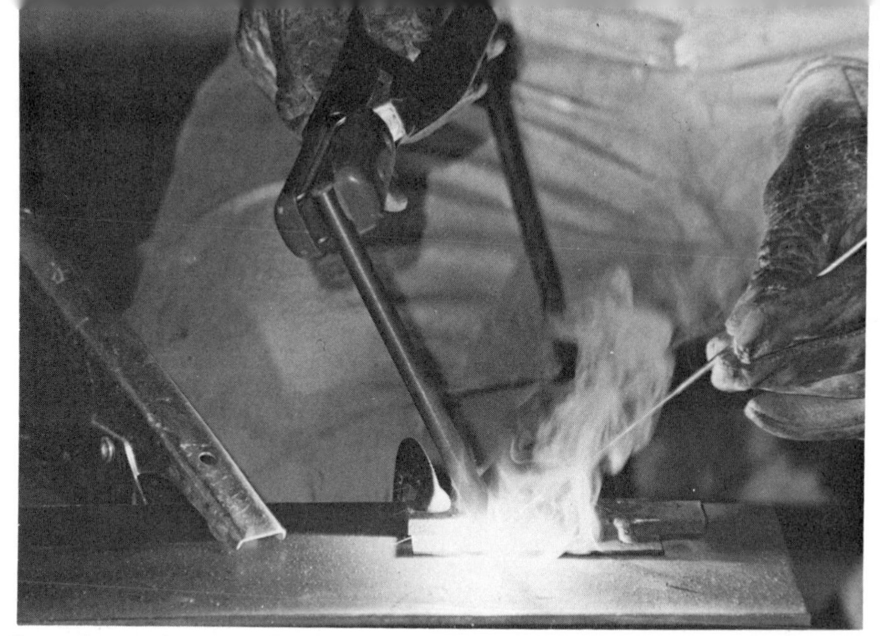

Grounding workpiece and using a single carbon in torch produces an almost instant spot-braze joint, above. Touch carbon to rod, not metal. Two samples, below, show brazing thick metal to thin, one with very wide gaps in fit-up bridged by braze metal.

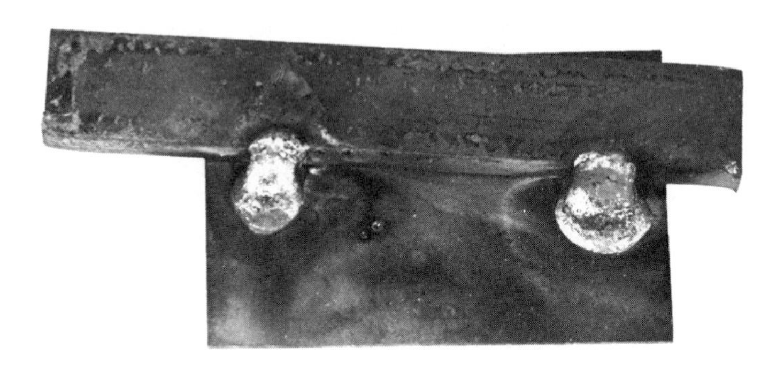

or paint the heat for a considerable area around the joint. If done properly this blends the heat into the mass of the work so there is less demarcation between hot and cold than in a weld. You can, of course, bring the entire workpiece to almost a brazing temperature and then arrange for slow cooling in lime or sand over many hours. Another procedure is to preheat, braze, and slowly cool while playing the torch over the work during an extended cooling period.

Remember, if you use the preheat and extended cooling with heat method, that your welder probably has a limited duty cycle. Holding an arc for a prolonged period may overheat the windings. My suggestion is to use a propane torch or oxyacetylene torch for preheating and cooling.

Single-carbon brazing. It is possible to do some rather unusual brazing and soldering by connecting the ground clamp to the work and supplying power to the lead to the fixed carbon of the torch. Instead of using two carbons you use only one, much as you set up for metallic arc welding. The difference is that the carbon will not melt and supply filler metal. You must hold the bronze filler rod in one hand and add metal as needed. This method has some advantages, since it permits a neat, very quick spot braze without introducing heat to a broad surrounding area. This helps enormously to reduce distortion in thin sheet metal. It also works almost magically when attaching a fairly heavy piece to lighter stock. An example would be joining something ¼″ thick to ¹⁄₁₆″ sheet. And if you have pieces that simply do not fit up well and have major gaps you'll find it easy to bridge the gaps quickly without dumping lots of filler into them.

The trick to this single-carbon brazing is to use a self-fluxing rod and place it firmly where you want the metal to go before you touch the carbon

Even thin sheet metal can be stitched together with a series of spot brazes using a single-carbon arc. Note that although spots are near edges there is no visible distortion.

to start the arc. Now, instead of striking the arc on the work metal, touch the carbon to the rod just above the contact with the work. You'll be surprised how quickly you get flowing metal and an almost instant spot braze. Most such spots are completed in a few seconds. You'll have to experiment a bit for just the right amperage. One suggestion: Sharpen the tip of the carbon before starting. Another: Use a DC carbon with straight polarity if you have DC.

SOLDERING. A similar technique but without an actual arc can make short work of relatively large soft soldering jobs. Again, clamp the ground to the workpiece which has been properly cleaned, fluxed, and clamped in position. With the welder tapped for 40 amps, at least as a trial, touch the carbon to the metal and hold it without drawing an arc. In a moment or two, touch the solder to the metal just behind the carbon. When it melts and flows into the joint, move the carbon along slowly and follow with the solder. Remember, if you are working with thin sheet metal, as you normally would be for soft soldering, lifting the carbon may draw an arc which might burn through. Either flip off the welder switch before lifting the carbon or flick the carbon up very quickly.

Soldering with a single carbon and the work grounded. Soft solder flows into joint without flame or fuss. Carbon is kept in constant contact with work to avoid arcing and possible burn-through. As always, gloves should be worn.

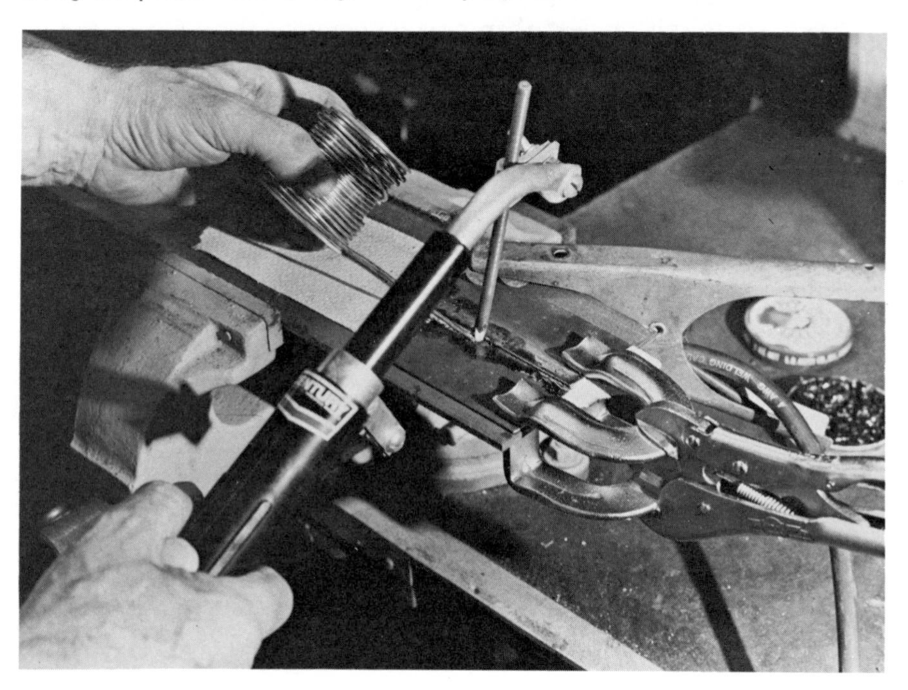

WELDING ALUMINUM. Ordinary soft aluminum alloys can be very successfully welded with a special flux-coated aluminum filler rod and the carbon-arc torch. But it's tricky, and I strongly advise lots of practice and developing a gambler's heart before trying a job on something valuable. Aluminum is hard for the beginner to weld for two reasons. The first is that it melts at a much lower temperature than steel and it does it without any really marked change in color or surface appearance. The result can be an instantaneous conversion of your workpiece into a molten blob. The second problem with aluminum is its formation of surface oxide as fast as you polish it off. The oxide prevents weld metal from adhering. For this reason, aluminum flame-welding procedures require a flux to remove the oxide and keep the air out until you can complete the weld.

One other point about carbon-arc welding of aluminum is that for practical reasons you want the workpiece flat and level in front of you. If you have a rounded part, be prepared to have a helper rotate it as you weld. In many cases it may be helpful to place the pieces to be welded over an iron or steel plate to absorb the heat and inhibit burn-through.

Many commercial aluminum welding rods are available. I've tried Lincoln's Aluminweld and Century's Arc Aluminum in ⅛″ and 3/32″ diameters. Both work; both are touchy for beginners. You will find it helps to look straight down between the carbons at the weld. Start by moving the arc up and down the joint for 3″ or 4″ to preheat the metal. Move the arc to the beginning of the weld and introduce the tip of the rod into the arc with your left hand. If the temperature is correct the coating should melt off and flow into the joint. Now let a drop of metal melt and fuse into the joint. Add metal by dipping the rod tip into the joint as you move along. The above instructions are basically Lincoln's for their Aluminweld. The technique is similar to gas welding with filler rod, but my experience is that after adding one or two drops of metal it pays to draw back the torch and let the metal cool before trying for more. Otherwise the heat tends to accumulate in the work and you suddenly have a burn-through.

Storing electrodes in a dry place is always good practice, but the heavily coated aluminum rods seem to be much more subject to deterioration from humidity than other rods. Even basement-shop humidity, especially in summer, will cause them to become wet and messy and destroy the coating. A good storage place for a few such rods, well wrapped, is on top of a refrigerator or upright freezer where a certain amount of heat is always present.

12 TIG (Tungsten–Inert Gas) Welding

TIG welding is probably the highest level of welding available to the ordinary home shop. It is also one of the most enjoyable because of the truly handsome joints you can produce. Although TIG welds can be made with DC it is much easier with AC and a high-frequency arc stabilizer. This device converts the electrical energy to the radio-frequency range. The power is not dangerous, at least no more than any other electrical source, but it has some unusual effects. One somewhat alarming effect is the tendency for small electrical discharges to emit from your hands and elbows to ground. Although these discharges are unlikely to cause serious burns and they cause no more than a tickling sensation, it is recommended that full protective clothing and gloves be used, as always. Full eye and face protection is necessary.

WHY HIGH FREQUENCY? The high-frequency current permits an arc to be struck without touching the electrode to the metal. Thus the high-frequency unit can be used for conventional metallic arc welding, and in some cases, such as welding with low-hydrogen rods, it is extremely helpful. In TIG welding, however, the tungsten electrode does not melt and does not touch the metal. Filler rod may or may not be needed, depending upon the joint. Two upturned edges, for example, may simply be melted together. When filler rod is used it is fed with the hand opposite from the torch in the manner of gas welding except that the angle between the rod and the metal is much shallower and in some cases the rod may actually be placed in the joint and melted together with the work metal. No flux is used on the work or the rod.

Two factors are involved here. First, it is claimed that the high frequency causes an agitation of the molten metal which tends to bring the oxides and contaminants to the surface so they don't interfere with adhesion. Secondly, the inert gas, argon, acts to keep out the air and prevent oxidation.

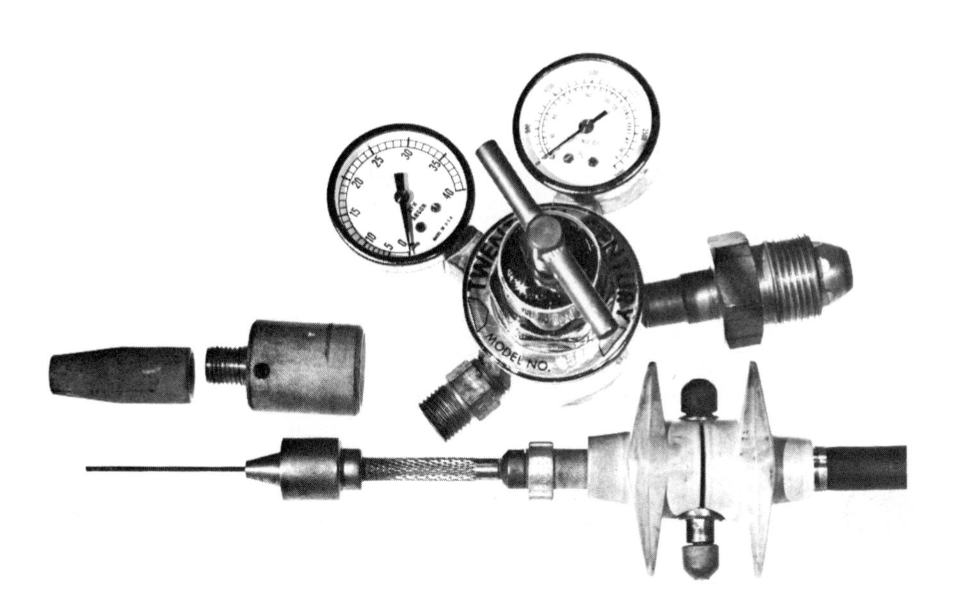

TIG torch and regulator/gages are needed, in addition to high-frequency converter, for TIG welding. Torch is partially disassembled to show ceramic nozzle, left; tungsten electrode clamped in chuck, below. Argon flow valve is on torch handle, right. Electrode holder grips knurled section of TIG torch.

HOW TIG WORKS. The TIG torch has a chuck much like an automatic pencil, and the tungsten electrode is secured in this chuck. A nozzlelike ceramic cup surrounds the electrode. Gas, almost always argon for home use, is fed from a high-pressure tank, through a regulator which reduces the flow to about 20 cubic feet per minute. A hose connects the regulator to the torch. Just before you start welding you press a button or valve on the torch to start the gas flowing. Argon is one-third heavier than air and also heavier than helium, which is sometimes used industrially. The argon settles and envelops the weld area to shut out oxygen, thus, in a way, acting much the same as the flux on welding rod or the burned gases of the oxyacetylene torch. Note that the argon gas is emitted from the ceramic torch nozzle around the center tungsten electrode.

When you first start TIG welding you'll almost certainly forget to turn on the gas before starting the arc. And you'll forget to hold the gas flow on the weld for a few seconds after you've completed a weld, or lift the torch away momentarily to allow cooling. In all cases you'll see instantly how molten metal, especially aluminum, oxidizes when air strikes it.

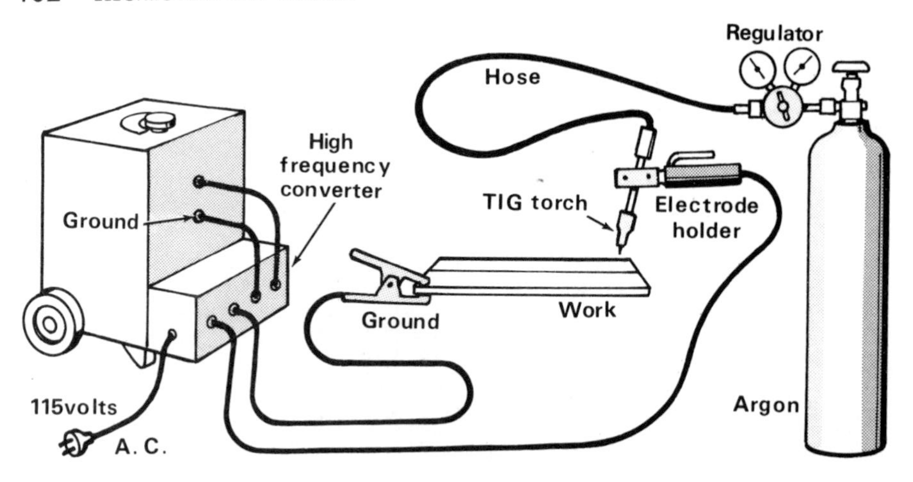

TIG welding connections can be quite complex, including water cooling for torch, but for home shop it's simple. High-frequency unit plugs into welder and welding cables plug into it. Argon hose feeds torch which is held in standard electrode holder. Main welder might be either 115-volt or 230-volt type, but extra 115-volt connection is needed for converter fan. Start/stop foot switch for welder, not shown, is very handy.

CONNECTING YOUR TIG TORCH. The actual connection of the high-frequency unit may vary slightly depending on the maker. In most cases it consists of nothing more than plugging the converter into the output and ground connections of the basic transformer welder and then plugging the ground clamp and electrode holder into the high-frequency unit. You will also need an extra 115-volt plug for the converter. The TIG torch is clamped in the electrode holder exactly like a conventional rod. The only difference is that you will have the trailing hose to the argon tank as well as the usual holder cord. Make sure both are free and don't interfere with your hand motion.

PREPARING THE WORK. Even though the argon acts to prevent oxidation and the high-frequency current tends to clean the weld, it is extremely important to provide meticulous cleanliness for TIG welding of aluminum. All grease, paint, dirt, or corrosion must be removed by suitable solvents or files followed by brushing with a stainless-steel wire brush. Never use abrasive cloth or paper, since they contaminate the metal. The stainless-steel wire brush must be perfectly clean and used for no other purpose than preparing aluminum for welding. Protect it from dust, welding spatter, and all the usual shop debris. This same cleanliness applies to aluminum filler rods. Don't let them become dirty or corroded.

As with all welding, the actual technique of TIG welding can only be learned by practicing. The usual torch angle is about 75°, about 15° off vertical, with the ceramic nozzle directed in the direction of movement.

The tungsten electrode may extend 1/16″ beyond the end of the cup, but my experience has been that this may be too much and it seems to work best when almost flush with the cup. For most home shops a thoriated tungsten electrode 1/16″ in diameter works fine at amperages between 70 and 150. Heavier and lighter electrodes are available from .020″ to 1/4″, but most home welding of aluminum will be below 150 amps and above 80 amps. The term "thoriated" tungsten refers to tungsten with an alloy to aid current-flow capacity and extend working life.

DIFFERENT TIG TECHNIQUES. After a bit of practice you'll find that there are several ways to handle TIG welding. As mentioned, you can weld some edge joints without filler. In other cases you can place the filler rod right in the joint. But one common procedure, which I call "go-and-back-up," is a good general-use method. Here, you establish a puddle, add filler rod to it, and bring the torch back slightly. Don't back up so far that the argon is withdrawn from the puddle. Now move ahead, again establish a puddle, add rod, and repeat the back-off action. This has the advantage of not letting the heat get ahead of the torch so the aluminum starts to collapse before you've reached the advance edge of the weld.

TIG-welded joint in aluminum, and section through it, showing full fusion of fillets. This would be a difficult weld without TIG.

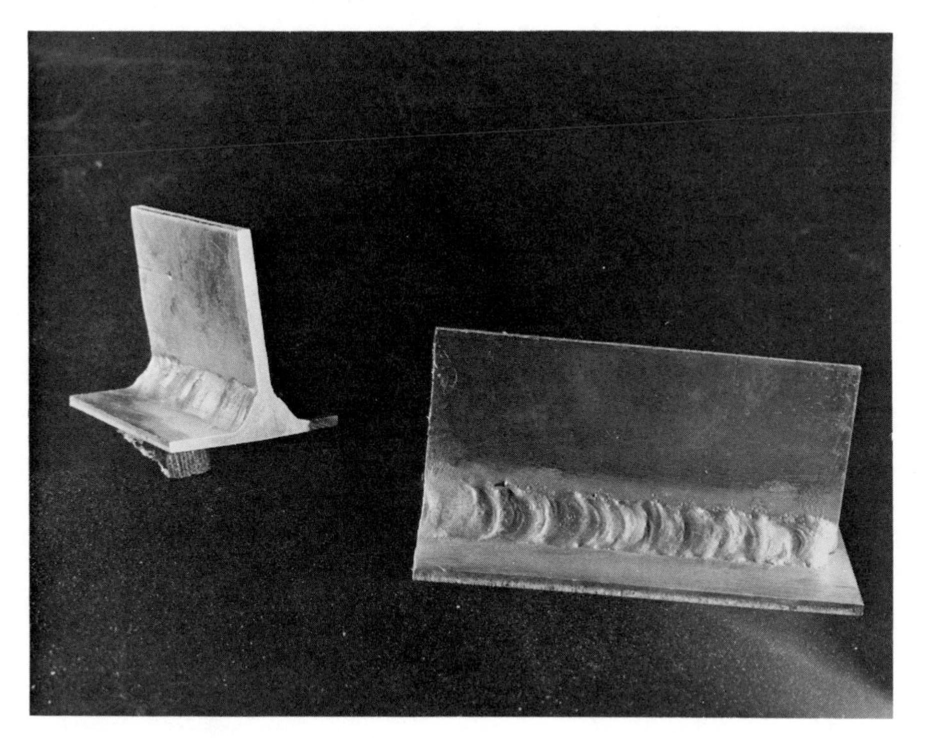

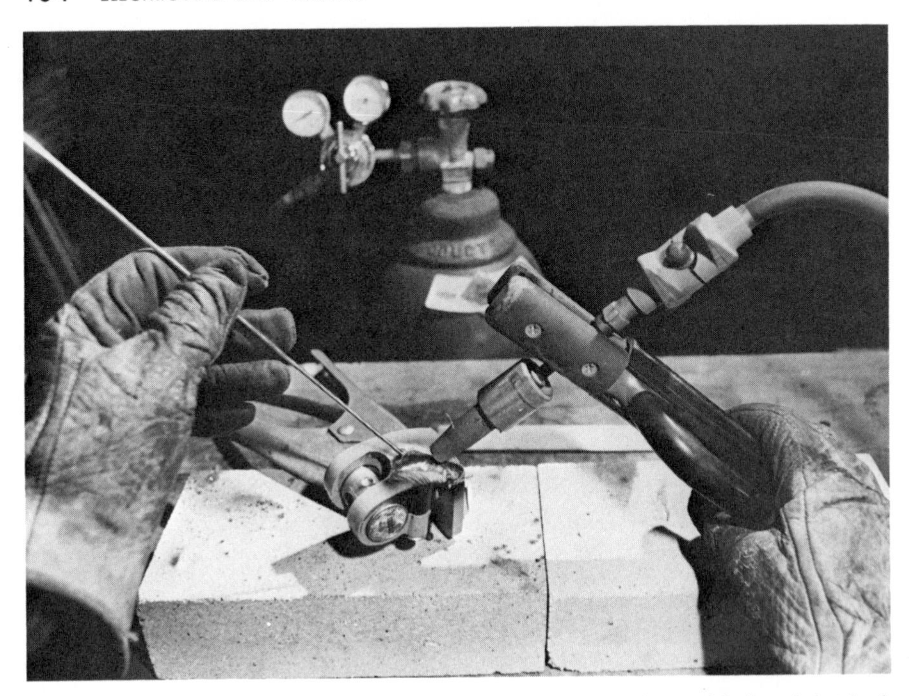

With TIG, aluminum castings such as this tripod head can be welded and finished off so repair is invisible. Unfortunately, all parts that look like aluminum are not aluminum. It's easy to get into trouble if part turns out to be die-cast metal.

One final point. As mentioned, the high frequency is in the radio-frequency spectrum and can raise havoc with radio and television if not controlled. Makers of such equipment are well aware of these problems and the proper control methods. Follow their instructions, and if your neighbor mentions that his favorite program is riddled with interference, play dumb and be sympathetic—and try to avoid TIG welding during his program.

13 Oxyacetylene Welding

Why gas welding? One very good answer is, "It's fun." With an oxyacetylene torch, or perhaps with miniature torches using MAPP gas and oxygen, even an unskilled person can assemble light art objects and make simple repairs. If arc welding is the champion in rough-and-ready structural work, then gas welding fits the light, delicate touch of the artist-craftsman in the home shop. And with oxyacetylene equipment you can flame-cut heavy metal to almost any shape and form.

An oxyacetylene welder uses a mixture of acetylene gas and commercially pure oxygen to produce a small, cone-shaped flame of about 6000° F. at the tip of the welding torch. In some respects this flame is easier to control than the arc. For one thing, you simply bring the flame to the work; you don't have to strike an arc. Since no ultraviolet hazard is involved and the light is not so intense, the protective goggles can have lighter lenses. You can see exactly where you're directing the heat and readily discern the color of the metal as it heats.

As discussed previously, it's nice to have both gas and electric welding equipment. If you must make a choice, do it on the basis of the type of work you want to do. If a metal-structure home-built airplane is your dream, gas welding will be a handy help, although you may also want TIG. If light metal furniture, auto-body customizing, or artifacts are your bents, by all means consider gas welding. One thing is almost certain—you'll enjoy doing it.

GAS AND ARC SIMILARITIES. You'll notice that the portion of this book devoted to gas welding is significantly shorter than the portion on arc welding. That's because so many aspects of gas welding are nearly identical to arc welding and repeating them would serve no purpose. The common features of both types of welding might be summed up as follows:

- Both depend upon fusing metal in a molten puddle.
- Both generally require filler metal.
- Both use butt, lap, tee, edge, and fillet welds.
- Both offer similar problems of expansion, contraction, heat warpage, and oxidation of unshielded hot metal.
- Both require fit-up, often edge beveling, clamping, position control, and heat shielding.
- Both require practice to develop reflex reactions of hand and eye to produce a high-quality weld.
- Both have certain basic hazards from heat, fire, sparks, and flying spatter. Electric welding has its special hazards because of the nature of electricity. Gas welding has its own particular hazards simply because of the nature of oxygen and acetylene gas, both of which are stored under high pressure and both of which can either burn furiously or cause other materials to burn.

GAS AND ARC DIFFERENCES. There are some major differences between oxyacetylene welding and arc welding which, although they may not affect the end result of joining metal, will affect your choice of welding method:

- Heavy structural parts are welded more quickly with arc, presuming the arc welder is the correct size for the job.
- Heavy parts require relatively long heating periods with gas; the cost of the gas or the amount left in the tank can be a factor.
- Extended heating with gas to achieve a puddle may cause damaging heat flow to other parts of an assembly.
- Gas welding may be the method of choice for difficult welds where metal thicknesses vary or corners are exposed. You can vary the heat by torch manipulation more readily than you can with arc.

ACETYLENE GAS. For the gas welder, acetylene gas is the true "wonder" fuel when mixed with oxygen. Not found in significant quantities in nature, acetylene made welding as we think of it today a practical art. Although a hydrocarbon, acetylene's molecular structure differs from the more familiar gases such as propane and methane. This peculiar structure, called a triple bond, actually releases heat during burning, and the resulting flame temperature exceeds those possible with other common gases.

For those of a historical bent, acetylene made by dripping water onto lumps of calcium carbide was long used as fuel for miner's lamps and for automobile headlights.

From the standpoint of the home-shop weldor there are some very important things to know about acetylene, since they influence how you handle it and particularly your own safety. Acetylene is unstable at pressures greater than the standard atmospheric pressure of about 15 pounds per square inch. This would make it impractical to store and transport in pressure cylinders except for another novel feature—it also has the ability to dissolve in acetone. The inner volume of the steel tank of acetylene that you bring into your shop is filled with a porous material saturated with acetone. On charging, the acetylene dissolves in the acetone and can be pressurized to about 250 pounds per square inch. We'll talk more about acetylene and the tanks that contain it later, but for now respect it as a unique energy gas with certain special requirements for storing and using it.

OXYGEN. The second important component of the oxyacetylene flame is pure, or almost pure, oxygen made by liquefying air and separating the nitrogen and other gases. Most of us are aware that oxidation, or the combining of a substance with oxygen, may proceed rather slowly, as with the rusting of steel; moderately fast, as in a wood fire; or very fast, as with the combustion of gasoline in an engine. Many of us have also seen the high school lab demonstration of thrusting a burning splinter into a beaker of oxygen. The splinter flares up wildly because it has a total oxygen atmosphere rather than the normal 21% we live in. And after burning the splinter it was part of the demonstration to heat some steel wool until it glowed and then drop it into the oxygen-filled beaker. The steel wool was immediately and dramatically consumed. This is the same principle as the oxyacetylene cutting torch, which heats a localized spot on the steel and then shoots a jet of oxygen onto the spot to burn a hole or a slit as the operator wishes.

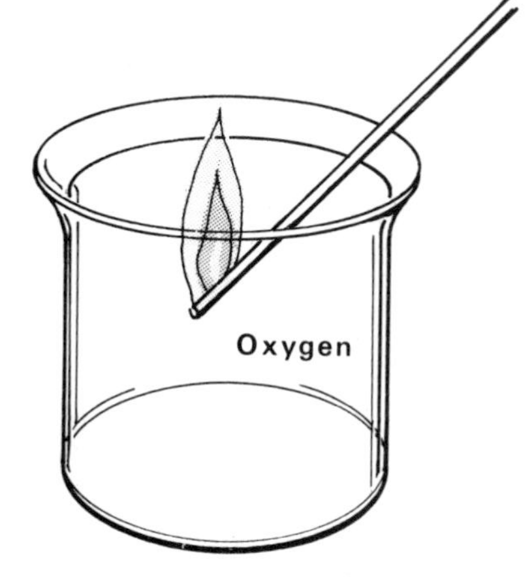

High school experiment shows ability of pure oxygen to bring a glowing splinter to flaming life. Cutting torch uses same principle. Most important, keep oil and inflammables away from oxygen.

I'm emphasizing this ability of oxygen to start and enhance combustion for safety reasons. Many oils and greases, even perspiration, can ignite spontaneously and burn furiously if welding oxygen contacts them. For this reason lubricants must *never* be used on welding fittings, hose couplings, torches, or regulators. Even more important, *never* direct a discharge of welding oxygen onto your clothes or person, or near a lighted pipe or cigarette. Even hair oil has been known to burst into flame; so has oily clothing. Always confine oxygen to the tip of the torch and then only when it is being used.

CUTTING AND BRAZING. In addition to its use for fusion welding the oxyacetylene flame is wonderfully handy for clean, accurate cutting of steel. Cast iron may also be flame-cut with special techniques, but it's not a home-shop job. You'll also find that you can use your welding torch for extremely delicate brazing, silver soldering, and (less easily) for aluminum welding. Heat for bending, annealing, tempering, and even crucible melting of small amounts of foundry metal are all easy with the versatile oxyacetylene flame.

14 Gas Welding Equipment

Scan the pages of the welding supply catalogues and you'll find a profusion of welding regulators, welding and cutting torches, and associated gear aimed primarily at the commercial and industrial user. There are many special needs and many quality levels and features offered with such equipment, but they seldom have much importance for the home-shop weldor.

What you want, especially in the beginning, is a complete outfit with regulators for acetylene and oxygen, hoses, a torch which can be fitted for either welding or cutting, several sizes of welding and cutting tips, goggles, and a spark lighter, all of this as part and parcel of a package matched together by one manufacturer. Although you may find an occasional bargain in used gas welding equipment, I strongly recommend against buying such. Regulators are delicate, and so are torches, tips, and the associated control valves. Even storage can damage them. Almost certainly a used outfit will have one or several components that someone has tinkered with or tried to repair. Hoses age and become damaged and leaky. Sometimes the threaded couplings which attach the regulators to the tanks get damaged so you can never get the necessary perfect seal against gas leaks.

In addition to making your learning process unnecessarily difficult, since you won't know if it's you or the equipment that's going wrong, there are genuine safety hazards involved. It's just better to stay away from used oxyacetylene equipment.

WHAT SIZE AND TYPE? You won't need the same capacity gas welding and cutting equipment for studio artifacts of light wire and sheet metal as you would in a junkyard cutting up old rail cars. My choice for general use in the home shop is somewhere in the range for welding up to ¼″ plate and cutting up to ½″ steel. Probably you'll seldom work with material

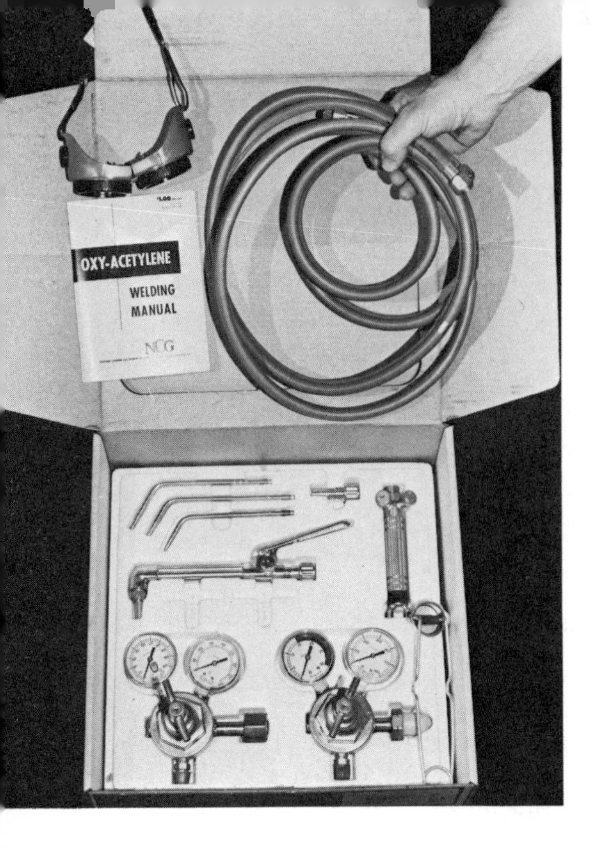

Light-duty oxyacetylene outfit has everything you need, except tanks, to get started in gas welding. It may be all you'll ever need unless you do heavy welding and cutting. Torch handle, upper right inside box, accepts welding or cutting attachments. Avoid buying used gas welding components. Delicate parts may be worn or corroded.

Medium-size torch, top, and smaller "aircraft" torch, below, are intended for heavier and lighter metals, respectively, but tend to overlap in home shop unless work is heavy. Note different locations for oxygen and acetylene throttles.

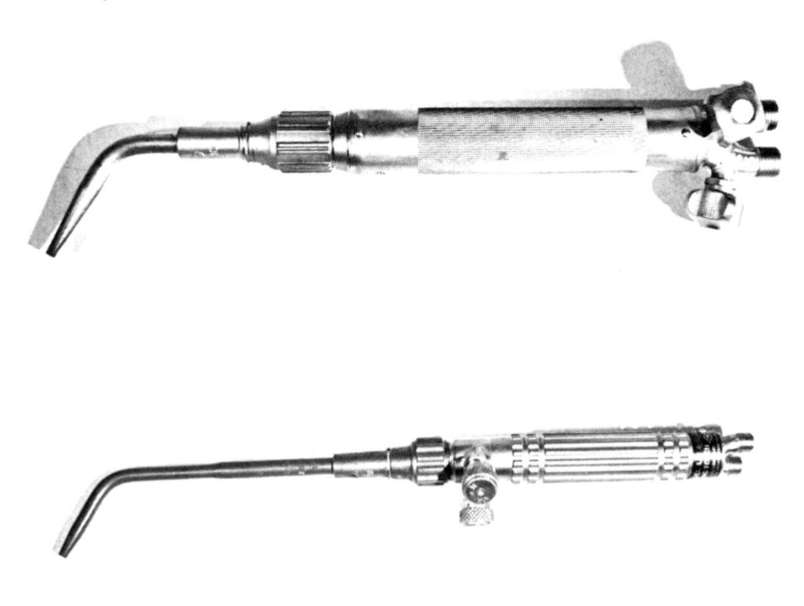

even that thick. I prefer what is sometimes called an "aircraft torch" for most welding.

At the bottom of a typical mail-order house's or big hardware store's lineup you'll find a combination welding and cutting torch, regulators, hoses, lighter, and goggles for about $100. In addition you may need to buy adapters to go between the regulators and whatever tanks you acquire or rent. And you may want a MAPP gas regulator, since MAPP gas works fine for cutting and costs less than acetylene. My own experience with such equipment is that although it cuts reasonably well and is more or less satisfactory for brazing, it is too unstable in gas control and adjustment for good gas welding.

A little more money, about $150, gets some much better equipment, with single-stage regulators, reverse-flow check valves, and heavier-duty hoses. For about $30 more you can get a torch with heavier working capacity, which you may not need, and two-stage regulation, which you probably don't need. The less expensive single-stage regulator will tend to drift off setting as you use up gas or as gas pressure changes. The two-stage regulators compensate for changes in supply pressures. This is important if you are doing production welding hour after hour. Since most

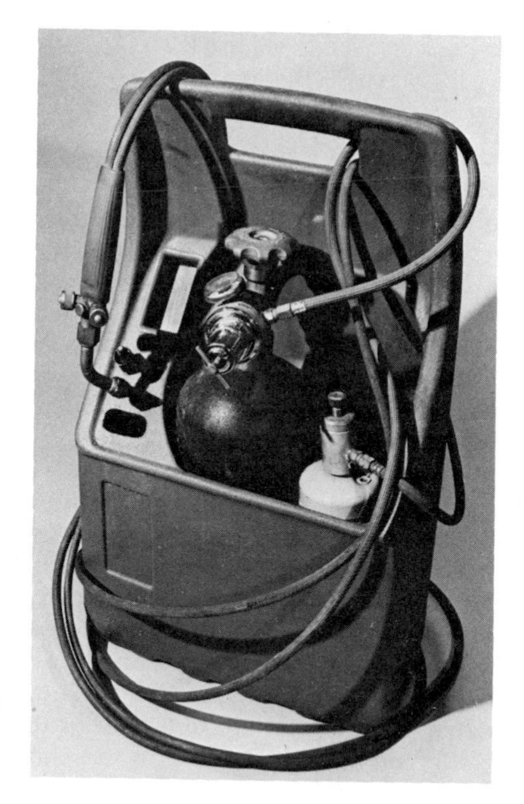

Small, high-pressure oxygen cylinder and MAPP gas in throwaway canister combine with simple regulators to give you a readily portable brazing, soldering, and cutting outfit. Do not, however, expect to do conventional fusion welds with MAPP.

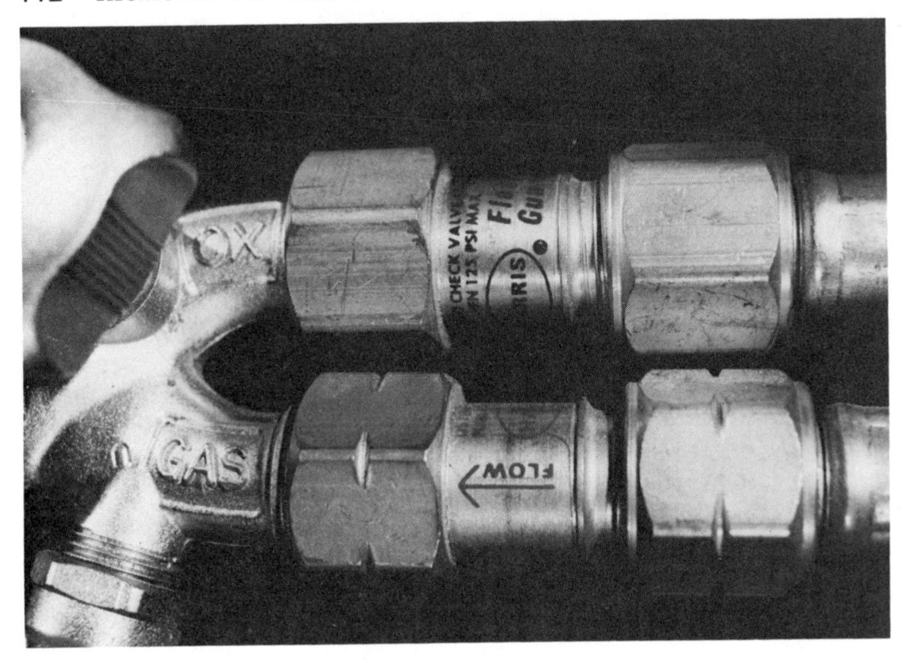

Two small reverse-flow valves between torch and hoses protect against flame flashback into hoses and regulators. You can aet by without them for years, but don't count on it.

home-shop weldors will probably use a tank of oxygen and acetylene off and on over a period of several months, the second stage doesn't really mean much.

SAFETY ITEMS. The reverse-flow check valves mentioned above are important and should be purchased separately if they are not part of the kit you buy. These valves act instantly to cut off flashbacks which occasionally start the gases burning inside the torch and may feed flame back into the hoses and regulators. Two of them, for acetylene and oxygen, cost about $10.

A spark lighter, or several of them, is an important although minor part of your welding equipment. Such lighters work by the flint-on-steel principle to produce a shower of sparks in a small metal pan. In use you direct the torch gases into the pan and squeeze the lighter legs. The flash or flame from the starting gas is somewhat confined in the pan rather than billowing onto your hand and arm as happens if you try to use a match or lighter for cigarettes. That's why I suggest buying several such lighters so you'll always (almost, anyway) be able to find one and not be tempted to use a match or other method to light your torch.

With this broad overscan of the equipment it's time to take a more detailed look at the individual components, since understanding them is much more important than with arc equipment. With a transformer arc welder you plug it in, switch it on, and it works or doesn't work. Gas equipment is capable of many shades of "not working," and some of them spell danger if you don't correct them.

THE WELDING TORCH. Although welding torches vary somewhat in detail design, the basic construction of the home-shop torch will be about as follows. You'll find two connections at the rear of the handle for the oxygen and acetylene hoses. The acetylene connection will always have a left-hand thread to mate with the red hose carrying combustible gas. The oxygen connection for the green hose will always have a right-hand thread. This prevents accidental interchange and applies throughout to all threaded couplings, including those for the regulators and the tanks.

Two thumbscrew valves will also be found on the handle for throttling and shutting off the flow of oxygen and acetylene. These valves should operate freely and smoothly, without wobble, and without being so loose that brushing them with your sleeve will cause changes in adjustment. Usually, the ease or difficulty of turning them can be changed by adjustment of the packing nut around the valve stem. The packing prevents gas from escaping and from that standpoint must be fairly tight. If acetylene should leak at this point it could ignite and burn your hand and arm.

The adjustment screws are an important point to check when buying

Schematic of typical torch shows feed of oxygen and acetylene through adjustment or "throttle" valves in handle, and mixing of gases in mixer just before tip. Smooth-operating, leak-free throttle valves are important for safety and essential for holding stable flame. Packing can be tightened, but significant wear or loose or damaged valve parts call for torch overhaul. *Diagram courtesy Union Carbide Corp.*

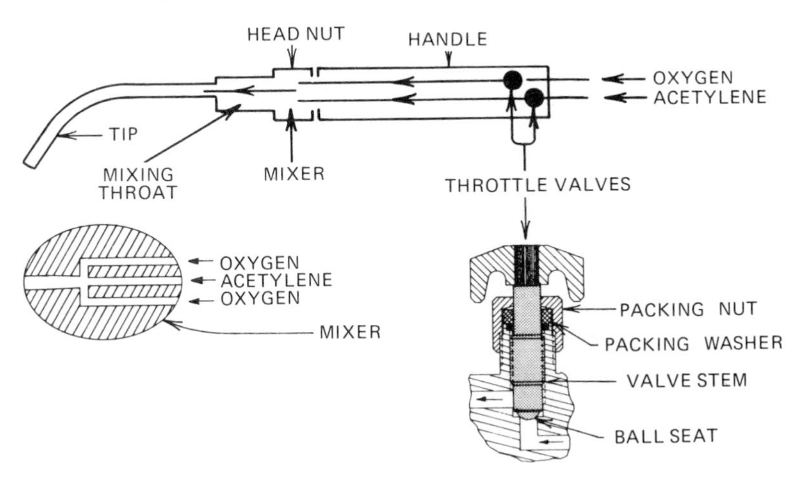

a torch and one of the weaknesses of very inexpensive outfits. As a check, try setting the torch for the desired neutral welding flame (see details page 138). Now try wobbling the valves between thumb and forefinger. If sideways pressure changes the flame, the valve is worn, damaged, or just too loose. Later, I'll tell you how to make another check with soapsuds.

Acetylene/oxygen mixing. Just ahead of the handle a separate section of the torch serves as a mixing chamber for the two gases. In the type of equipment used in home shops the oxygen and acetylene delivery pressures are usually set about equal. The mixer is sized in its internal passages to match the flow needed for the size tips intended for use with a given torch. Thus a small torch will not handle a large welding tip, and vice versa.

Welding tips. The final mixing of the gases and control of the flame size is a function of the tip, which is secured to the torch and mixer head assembly. These tips are very carefully made of copper alloy to conduct heat away from the immediate flame area. Their internal shape, the hole in the end, and general condition have a considerable effect on performance. All too often a weldor will casually toss a torch onto a bench or remove a tip and toss it into a box of mixed tools. Since the metal is quite soft such treatment usually damages the delicate gas passage outlet. Once

Welding torch tips for small torch, top, and larger torch, below, are not interchangeable. Lower tip with large nozzle is special for oxygen/propane heating, more economical than oxyacetylene.

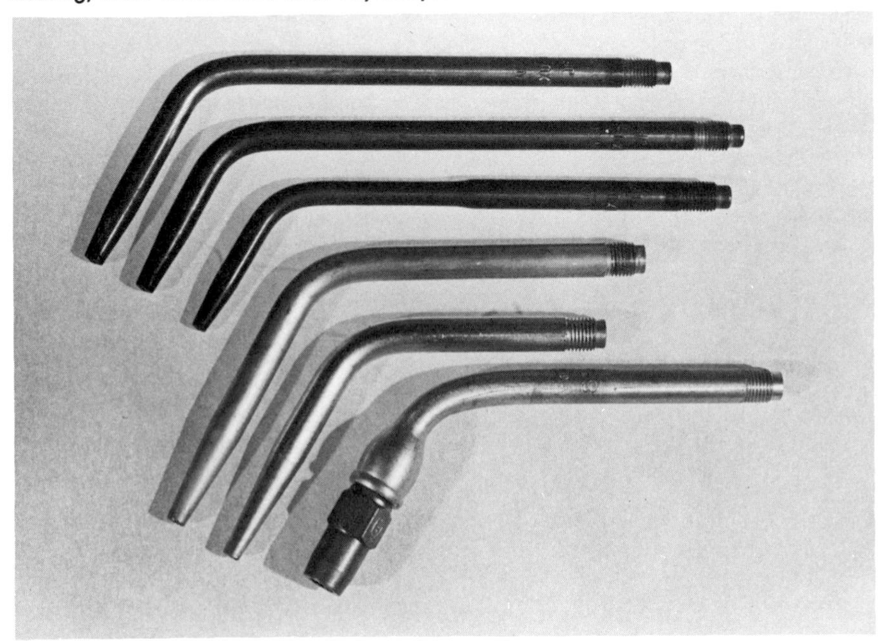

Casual use of wire and drills in attempts to clean the small orifices in torch tips and nozzles almost always makes matters worse. Tip-cleaner tool, left, has a large selection of cleaning wires with spiral cutting edges. Careful matching of wire to orifice size and gentle use will usually get a clogged tip working again.

a tip has suffered even minor damage it can seldom be brought back to perfect performance even with the insertion of a properly sized tip-cleaning tool.

Tip sizes. It would be nice if the tips and numbers identifying their sizes were standardized throughout the industry. Unfortunately, the sizes, numbers, and sometimes letter identifications do vary from one maker to another. Very often the threaded ends of the tips which screw into the torch are also different. In some cases the tip numbers are claimed to represent the cubic feet per minute of gas that they will deliver under the recommended pressure. In other cases the number corresponds to the number drill size of the hole. In still other cases you can only guess at the meanings of the several letters and numbers stamped on the tip. Probably some are only part numbers with no reference at all to operating characteristics. Once again, that's just another reason to buy a package kit new and stick with tips and replacement components from that manufacturer.

Cutting torches. Although many commercial users have cutting torches made only for cutting, and some are elaborate indeed, the average home-shop outfit consists of a handle with two throttle valves to which either a welding head or cutting attachment may be affixed with a few turns of a

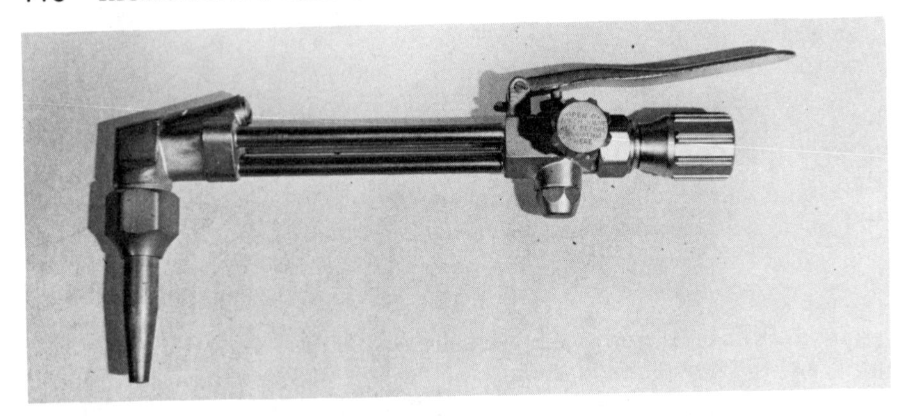

Cutting attachment for small torch secures to basic torch handle. Full oxygen pressure is released from central orifice in nozzle when cutting lever, top right, is pressed. Oxygen for preheat orifices is adjusted by small knob, below.

retaining nut. The difference between the welding attachment and the same unit with a cutting attachment is that the latter provides a second oxygen delivery passage and a second oxygen throttle valve. In addition, another control or triggerlike valve must be provided to enable you to release the stream of cutting oxygen at the right time and stop it at the end of the cut. Later, we'll talk about how these additional controls are used.

Cutting nozzles. The tip used for cutting is often called a cutting "nozzle" and instead of a single outlet for the mixed fuel gases there will usually be either four or six such outlets surrounding another single central outlet.

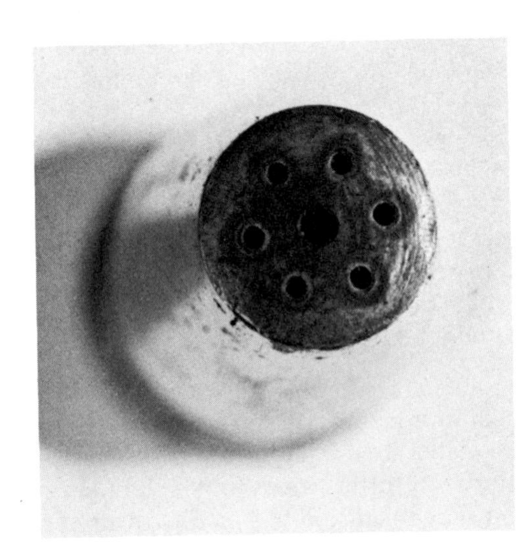

Close-up of orifices in cutting nozzle shows central orifice for cutting oxygen and six surrounding orifices for preheat flames. This nozzle shows some small deposits from slag spattering.

When you first start the cutting operation the multiple outer outlets produce a ring of small flames for heating the metal. As soon as you've preheated a spot on the steel to a red, called a kindling, temperature, an easy pressure on the oxygen trigger releases a blast of higher-pressure oxygen through the central hole. The oxygen immediately starts the preheated metal burning and makes a hole or cut depending on torch movement.

Thus, the usual adjustment knob for oxygen on the torch handle is adjusted wide open and the preheat-flame oxygen supply is adjusted by the knob on the cutting head. Only when the trigger is operated does the full oxygen pressure, usually 25 to 65 psi (pounds per square inch), come into play, and then only through the central nozzle hole.

PRESSURE REGULATORS. By far the most delicate and important parts of any oxyacetylene outfit are the pressure regulators and gages which mount on the tanks. Since a fresh tank of oxygen has about 2200 psi internal pressure, and a fresh tank of acetylene up to 250 psi, it follows that the regulators have quite a task to do in reducing the pressure on each to a smooth, steady, flowing 1 to 5 psi at the torch for welding. In theory, such a regulator might be constructed by allowing tank pressure to enter, pass through a variable orifice controlled by a tapered valve under a spring pressure set to balance the tank pressure at a selected value. In practice, a great deal of research and engineering have gone into working out the details of the rather complex and delicate inner workings of the regulator valves. The basic elements, however, are much as described, with the addition of a diaphragm to seal and prevent leakage. If you examine a typical regulator externally you should be able to look up into the tank inlet connection and see some sort of porous filter to prevent even minute bits of dirt from entering the precision inner valve. Moreover, you'll note that the inlet connection surfaces are nicely finished to provide a gastight seal to the tank fitting. Dents or scars in the metal will probably cause leaks.

You'll also see an outlet connection for the hose and two gages. One gage measures the high-pressure gas as it comes from the tank. It tells you, primarily, how much is left in the tank. The other gage tells you the pressure you're delivering to the hose and torch. You must make that adjustment before welding or cutting by turning in the faucetlike handle in the center of the regulator body. Turning the handle in brings spring pressure to bear against the internal working parts. Increasing the number of inward turns increases the pressure and allows the gas to flow to the torch. Backing out the handle reduces the pressure and ultimately shuts off the flow.

When you're finished welding you should always close the tank valves, open the torch to bleed off the trapped gases, and then back both regulator handles out until they feel perfectly free. If you back a handle all the way out, no harm is done. Put it back.

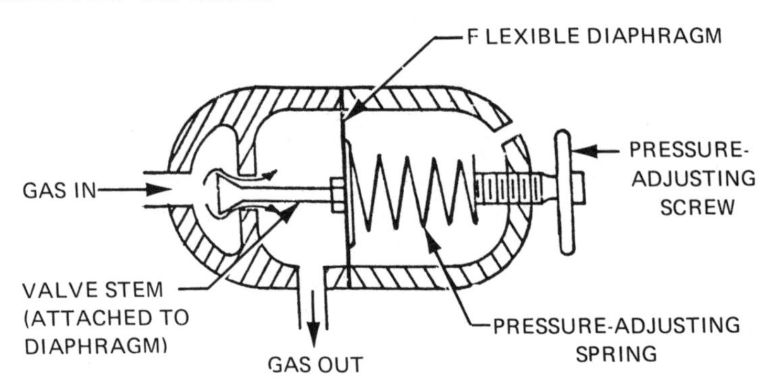

In principle, pressure regulator is a simple device that balances gas pressure with spring pressure to control a tapered valve in an orifice. Diaphragm acts as seal. *Diagram courtesy Union Carbide Corp.*

There is not much you can do about servicing a regulator. If one ceases to operate properly, or if—I hope not—you tip over a tank and damage one, forget all thoughts of home repair. Return it to the dealer for factory rebuild.

In actuality, regulators are very precisely built and contain a number of relatively delicate parts which can stick, corrode, or be impaired by dirt. Filter in inlet is very important. *Diagram courtesy Union Carbide Corp.*

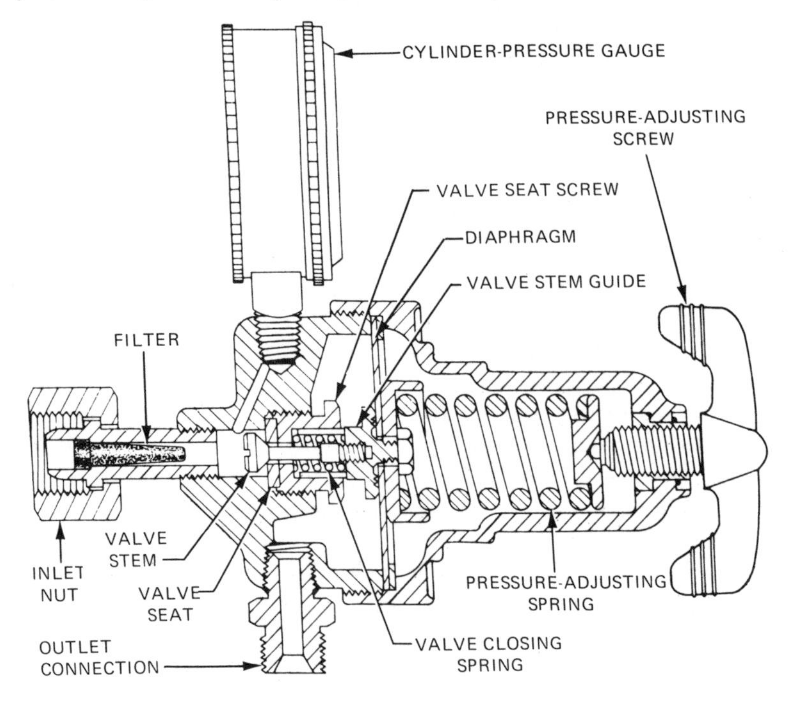

Gages on regulators tell you the pressure in the tank and what pressure you're feeding to torch. Here, lower gage shows fresh acetylene tank has about 250 psi. pressure. Turning in regulator handle has adjusted delivery pressure to torch to about 4 psi.

Regulator gages. If the gages on your welding outfit appear to be pretty much standard equipment such as you've seen on your car's instrument panel, your well pump, or your air compressor, they are—except that they handle much higher pressure and can be damaged by careless use. The gages are basically of the Bourdon-tube type. Inside there's an almost flat metal tube, sealed at one end and open to gas pressure at the other. The tube is curled into a semicircle. As gas under pressure enters the tube it tries to straighten out, and the small mechanical movement is transmitted through gears to rotate the needle.

Gage damage. How can you damage gages? Mostly by opening the gas cylinder valves too abruptly and while having the pressure-adjusting screws turned in rather than backed out. Opening the valves slowly allows the high-pressure gas to ease into the rather delicate tubes, and with the regulator backed out the sudden pressure is not applied to the low-pressure outlet gage before the regulator has a chance to control it.

One other rather curious form of gage damage called "burnout" is described by makers of such equipment. This applies only to the oxygen regulator gage, and to understand it you have to hark back to what I said about oxygen being able to start materials burning spontaneously. If the regulator filter is damaged or missing, and if any of a variety of shop dusts, dirt, paint particles, or the like have entered the Bourdon tube, it's possible, if you open the tank valve too quickly, for the sudden rush of gas to compress, and thereby heat, the oxygen trapped in the tube. This combination of heat and oxygen can cause an explosion that damages the regulator and maybe you. The moral of this is never to remove the regulator inlet filter or operate without it and always to cap the regulator inlet and hose connections during storage.

A final point about gages. Never position yourself directly over or in front of them when opening a cylinder valve. Mechanical failure or unsuspected damage could result in sudden failure and bits of glass and regulator parts striking you in the face.

Safety features. As a protective measure, all regulators for oxygen and argon and other noncombustible gases are equipped with blowout discs or similar relief provisions which will fail automatically in the event of excess pressure on the outlet side of the regulator. Combustible gases for uses other than welding do not have such reliefs built into the regulators, but for welding they must have a safety relief and vent incorporated. Anytime you hear a venting sound or smell leaking acetylene, it's time to close the tank valve and find out what's wrong.

CYLINDER GAS CONNECTIONS. We'll discuss the actual steps in connecting your regulators to the tanks later, but before reaching that stage it's important to understand some practical problems which you may encounter and which you can avoid if you know about them. The rounded-end oxygen connection for oxygen tanks is standardized and presents no real problem, since all standard regulators fit all standard oxygen tanks. The exception is the small screw-in oxygen tank used with certain portable

Install your regulators gently to avoid battering or scarring seating surfaces, which must seal against high pressure. Slide the coupling into place and move it slightly until it feels seated and aligned before trying to turn down coupling nut.

MAPP/oxygen units. Such tanks contain only 400 psi when full; the use of such equipment will be discussed later.

Acetylene cylinder and regulator connections are standardized, but, unfortunately, there are several standard types which cannot be interchanged. Thus if you buy a welding outfit you may find that your acetylene regulator cannot be attached to your tank without an intermediate component called an adaptor. The differences reflect acetylene tanks from different manufacturers and distributors. In case of mismatch it may just be easier to buy your acetylene from a different supplier with tanks you can match. In many home shops, however, weldors prefer to use the small bottle-shaped acetylene tanks, and with these you'll almost certainly need an adapter.

Another word of warning is in order. The commonly used intermediate-size acetylene tank has a rolled collar around the top. This is handy for carrying and protects the gages. For reasons unknown, some makers of regulators provide a stem between the regulator and the connection which is a fraction too short. Although there is no problem in starting the coupling thread and drawing it down, you may find that the side of the regulator body contacts the rolled collar just as you snug down the connection. This, of course, flexes the connection stem and regulator and strains it badly. A dangerous crack or leak may result. The only solution is a different regulator or an adapter to raise the body clear of the collar.

Hoses and connections. The rubber hoses which carry oxygen and acetylene from your regulators to the torch are made specifically for the gases, pressures, and service involved. Never try to substitute hose from a home paint sprayer or the like. For most home-shop use the hose length seldom needs to be over 12′ to 15′, and you will seldom need hose with an inner diameter over 3/16″. The heavier the hose the bulkier and more bothersome it is to handle with the torch, especially on delicate work. On the other hand, if you anticipate needing a hose over 25′ long and dragging it around on a construction job, for example, you may want to use 1/4″-inner-diameter hose to reduce the pressure drop between the regulator and the torch. If the hose will be exposed to oil and grease such as on a shop floor, be certain to get an oil-resistant hose.

The hose end couplings are another standardized product made specifically for coupling to regulators and torches. There are three basic sizes, A, B, and C. There are a number of variations in the method of inserting and securing the fitting in the hose ends. Most shop welding equipment will have a B-size connection, but in the case of smaller, hobby-type torches you may find the A-size used. Before buying hoses or connections determine which you have or take the old hose along to match it.

GAS WELDING GOGGLES. Although gas welding presents less ultraviolet hazard than arc welding, it does produce infrared radiation and

Small gas welding goggles, left, are fine if you don't wear eyeglasses. Larger ones, right, fit over glasses. The side covers are just as important as the lenses for keeping spatter out of your eyes. Never use sunglasses or other makeshift substitutes!

quite intense visible light. Welding goggles are essential! *Never* try to get by with dark glasses, sunglasses, or the like. All gas weldors get an occasional "pop" in a moment of carelessness when they dip the tip into the molten metal, and there are always minute sparks and globules of hot metal and scale flying about. If you doubt this, examine a pair of welding goggles which have been used a bit. The pits and lumps on the glass surfaces speak for themselves. This makes it vital that the goggles be equipped with snugly fitting side covers to prevent such material from intruding behind them and getting into your eyes. Take no chances, even for a moment.

In gas welding, perhaps even more than arc welding, it's very important to be able to see the puddle and fusion action clearly. If you must wear glasses to read and do close work, you'll certainly need them for welding. Since nearly every welding kit comes with little round goggles, or worse, which simply do not fit properly over glasses, you should purchase another pair right at the beginning which will cover your glasses. Even so, if you wear bifocals or trifocals you may still have trouble. That's why I had a pair of prescription close-up glasses made for wearing under goggles and arc welding helmets.

Welding goggle lenses are replaceable and should be chosen to suit your work. Small, typical home-shop jobs can be safely handled with a #4 lens; for heavier work, or if your eyes feel uncomfortable after welding, go to a #5 or #6 lens.

GAS CYLINDERS. In spite of their often battered and discolored appearance, gas cylinders are among the most carefully built and thoroughly inspected and reinspected products in our society. Oxygen cylinders are under the jurisdiction of the U.S. Department of Transportation (at one time the ICC) and must be stamped according to the specification of the cylinder, the filling pressure, the date of manufacture, and the date of each retesting. In addition, you'll find the maker's size and serial number or trademark.

As an example, the oxygen tank in my shop at the moment has an ICC3A 2015 mark. This means it is a 3A specification and should be filled to a maximum of 2015 psi. Another series of numbers tells me that it was manufactured in 1947 and tested a number of times over the years, the last time in 1977. Obviously it's a fairly old tank. I can also tell by the fresh thread sealing tape at the outlet valve that the valve was recently replaced.

For these reasons, you do not actually "buy" the larger oxygen tanks used in welding. This does not apply to the very small tanks used with MAPP/oxygen units. When you first start gas welding you'll probably go to the supply shop and come home with full tanks of oxygen and acetylene. You'll also make a deposit on the tanks. If you use all the gas within a month and go in for replacement, you'll pay only for the gas. If you take the tank back partially full, you'll still pay full price. There is no practical way for the dealer to measure the residual gas. If, as most of us do, you

Oxygen cylinder tanks will show ICC, or, recently, DOT specification number, filling pressure, and other proprietary marks meaningful to maker and service agency. Opposite side of this tank shows date of manufacture and dates of subsequent retesting.

Open handwheel valve of oxygen cylinder very slowly to reduce the impact of high-pressure gas on regulator and gage. Two-handed grip helps. Note that regulator handle is fully backed off.

keep the tank for several months, you will be billed each month for tank rental.

Eventually this will probably become annoying enough to make you check into a permanent tank lease. With such an agreement, actually a fairly imposing legal document, you pay once for a lifetime and after that pay only for the gases. This arrangement apparently varies over the nation, and the only way you can know what's best is to shop around locally.

Fire protection. When you bring an oxygen cylinder into your shop you must never forget that it has an internal pressure of over 2000 psi. Since the tanks are tested to a much higher pressure, this presents no hazard under normal conditions. A fire, however, might heat the tank and increase the pressure to the bursting point unless some relief was provided. Therefore, all oxygen tanks have a safety disc in the side of the outlet valve which will rupture well before excess pressure develops. This is better than a bomblike explosion, but it may also feed pure oxygen to the fire and increase its intensity enormously. For such reasons it is wise to keep your welding equipment well away from your furnace or any other potential source of heat.

Oxygen cylinder valves. When you receive your oxygen cylinder it will have a heavy metal protective cap screwed on the top. Remove the cap and directly underneath you'll find a large brass valve with a small handwheel for opening and closing. You should know two things about this valve. First, it must work smoothly but never be oiled or lubricated. Second, when you open the valve—slowly, to protect the regulator and gages—you must open it all the way until you feel it seat firmly at the top. That's because the valve has a seal at the top as well as the bottom to prevent leaks while you are using oxygen.

My practice is to mix some dishwasher soap with a little water to make a suds whenever I bring home a new tank of oxygen or acetylene. I brush this solution around the point where the valve threads into the tank and around the valve stem. Occasionally you'll find a leaker and have to exchange the tank.

Acetylene cylinders. As described earlier, the acetylene cylinder, which is just as carefully made as the oxygen cylinder, is filled with a porous material saturated with acetone. The gas is dissolved in the acetone. Although the pressure in the tank is about 220 to 250 psi at room temperature, there is always the possibility of overheating from fire. A sudden venting of a combustible gas into a fire is obviously unwanted, and acetylene cylinders are fitted with temperature-sensitive plugs which will melt out and allow a gradual release of the gas. One very important thing to know is that the melting temperature of the plug is about the same as that of boiling water. Clearly, you shouldn't store an acetylene tank near a furnace or other source of heat. If you happen to store a tank outside and

Acetylene tank requires a special tee-handle wrench. One turn is almost always enough. Leave wrench engaged for quick use if an emergency arises.

it collects ice or snow on top, *never* use boiling water or steam to remove the ice.

Because the cylinder has liquid acetone inside, it is important that acetylene tanks be stored upright. If, as most of us do, you bring home such a tank in your car trunk placed on its side, allow it to stand vertically for a while before connecting your equipment and trying to weld. Never use an acetylene tank while it's resting on its side.

Unlike the valve on the oxygen tank, the valve on most acetylene tanks has a short, square stem projecting. Opening this valve requires a short tee-handle wrench, which your dealer will give you when you get the tank. There is seldom any reason to open the valve over one turn in home shops. Never exceed one and a half turns. Always leave the wrench in place for a quick shutoff in an emergency.

Tank sizes. Unless you operate a commercial welding shop or have very large projects the sizes of oxygen and acetylene tanks you use will probably depend more on convenient portability than anything else. Surprisingly, there is no standard designation for tank sizes. An answer to my query about this to Linde Division of Union Carbide was, "Our cylinder size designations of Q and WQ have . . . no meaning," and "Each manufacturer applies his own designations and even cylinder colors." Even so, you'll find that each manufacturer offers tanks of approximately the same size regardless of how they designate them.

My recommendation for the home shop is the size Linde calls Q for acetylene and WQ for oxygen. The acetylene cylinder is about 25″ high, and the oxygen cylinder about 30″. Thus they fit together well on a cart without the gages bumping each other. More important, although fairly heavy, neither is too heavy to carry in and out of a basement shop or lift in and out of a car trunk. There are at least two smaller sizes of acetylene tanks used primarily by plumbers and air-conditioning persons carrying their equipment about the job sites with them. These are fine if you're doing light, art-type welding in a studio, but they keep you running for refills if your jobs are a little heavier. Also, there are many much larger tanks that require hoisting equipment.

As a final note on gas cylinders: They are not intended to be used as rollers to move heavy equipment, as anvils for hammering or shaping metal, nor as weights and props in the shop. Never strike an arc on one or direct a torch flame against it.

NON-ACETYLENE MINI-TORCHES. A few years ago a new heating gas called MAPP (for stabilized methylacetylene and propadiene) was introduced, and its excellent heating qualities plus the fact that it can be compressed (unlike acetylene) and sold in small, portable canister tanks like propane made it instantly popular. It was also overadvertised and misrepresented. In addition to being supplied for some new general-pur-

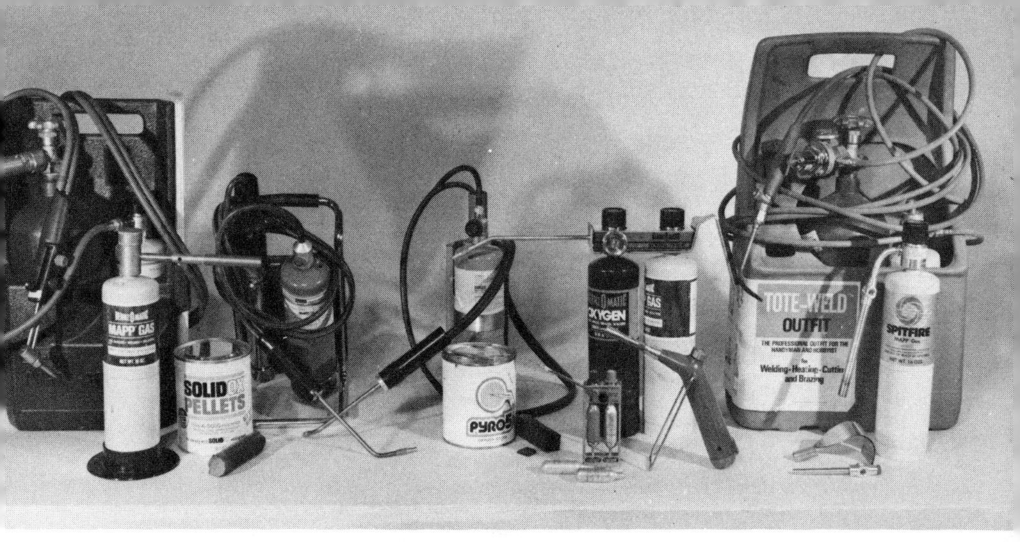

Mini-torches range from larger MAPP/oxygen brazing and cutting outfits using high-pressure oxygen bottle, to MAPP torches using low-pressure oxygen canisters or solid fuel which you ignite to produce brief flow of oxygen. Smallest is butane/oxygen hobby torch, right center. All, including MAPP/air, are good for heating, some for light brazing, none for really satisfactory fusion welding.

pose air/gas heating torches formerly confined to propane, it offered excellent flame-cutting ability when combined with oxygen. MAPP can also do many light brazing jobs.

A number of special torches were devised for MAPP, and some of them were advertised as "welding equipment." Others, more realistically, were offered only for cutting and brazing. One of the latter comes with a small high-pressure oxygen cylinder and a MAPP tank with regulator and a torch all neatly packaged in a heavy plastic "tote" box. This unit has proved extremely useful for workmen or home-shop users needing a portable cutting torch of high heat for silver soldering or brazing.

Other MAPP outfits were offered with canisters into which you inserted a stick of special chemical which released oxygen when ignited. Another rig came with a screw-in, 1.1-cubic-foot, low-pressure oxygen canister.

To be honest, none of these did a good job of fusion welding on steel. Although the temperature attainable with MAPP and oxygen was claimed to be only about 200° or less lower than acetylene and oxygen, the heat distribution in the flame is different and getting a puddle of molten metal is very difficult. If you have seen artwork of wire and light sheet metal made with such equipment you may have noticed that the so-called welds had an odd, porous appearance quite different from a cleanly executed fusion weld with oxyacetylene. Nevertheless, many enthusiasts adopted these torches because of their low initial cost and because of their convenience for artifacts where strength isn't very important.

My personal trials with them for ordinary home-shop welding left me with no confidence in their utility, or even their long-run economy. Simple welding jobs, such as repairing cracks or broken joints on the light tubing

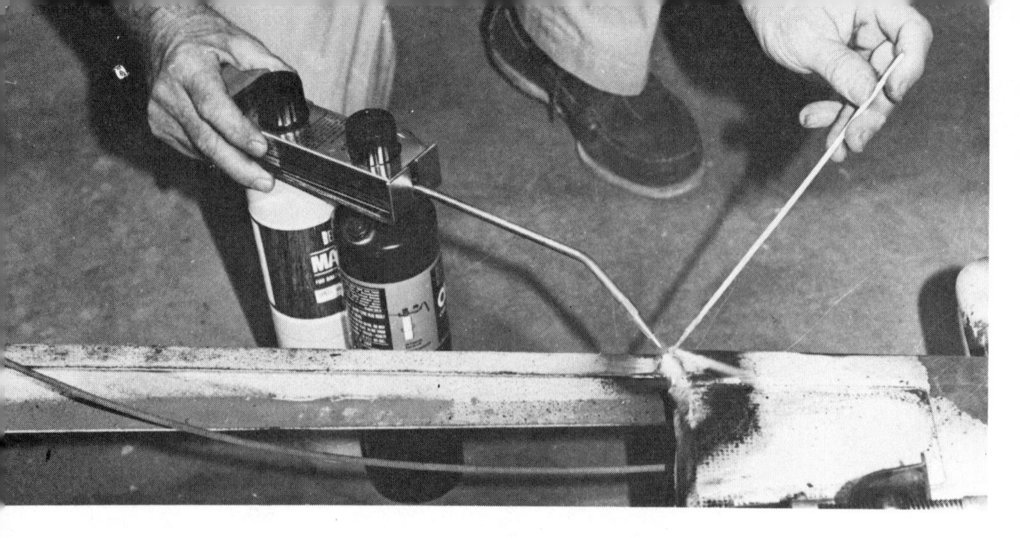

Brazing with pre-fluxed rod worked better on this repair than attempts to fusion-weld with MAPP and canister oxygen. Small, low-pressure oxygen supplv doesn't last very long.

Brazing went fairly well with this portable high-pressure MAPP/oxygen rig, but torch adjustment was loose and unstable.

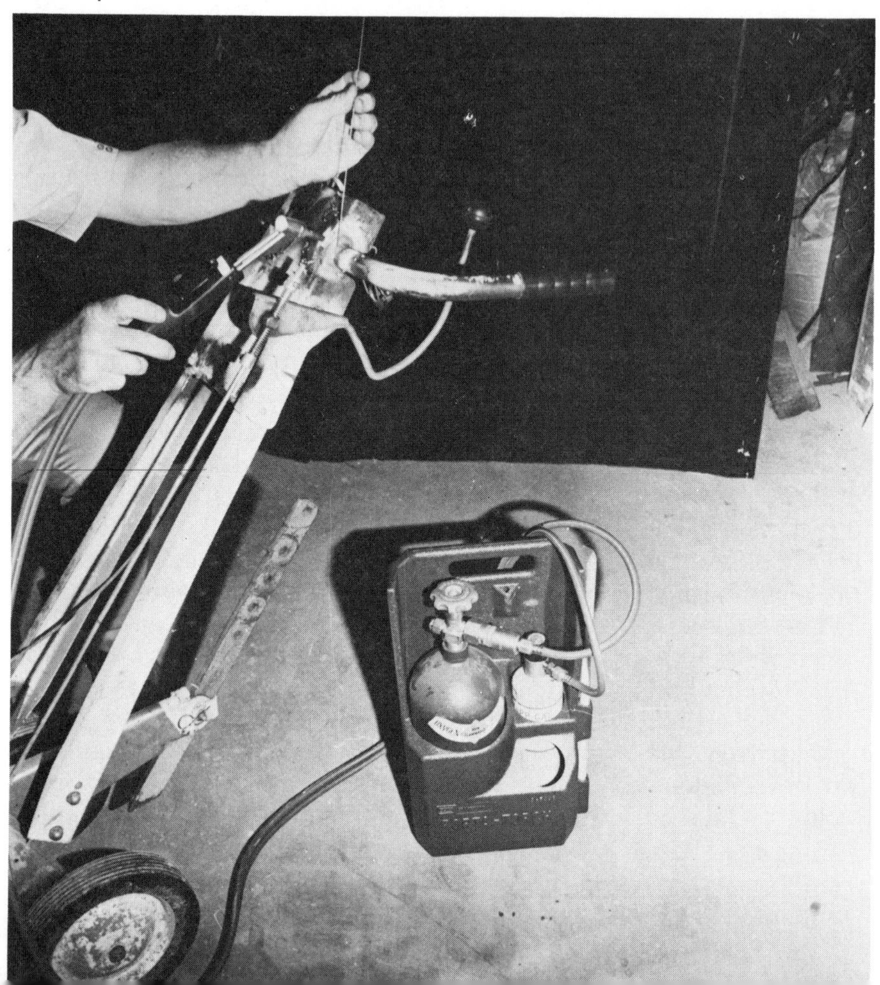

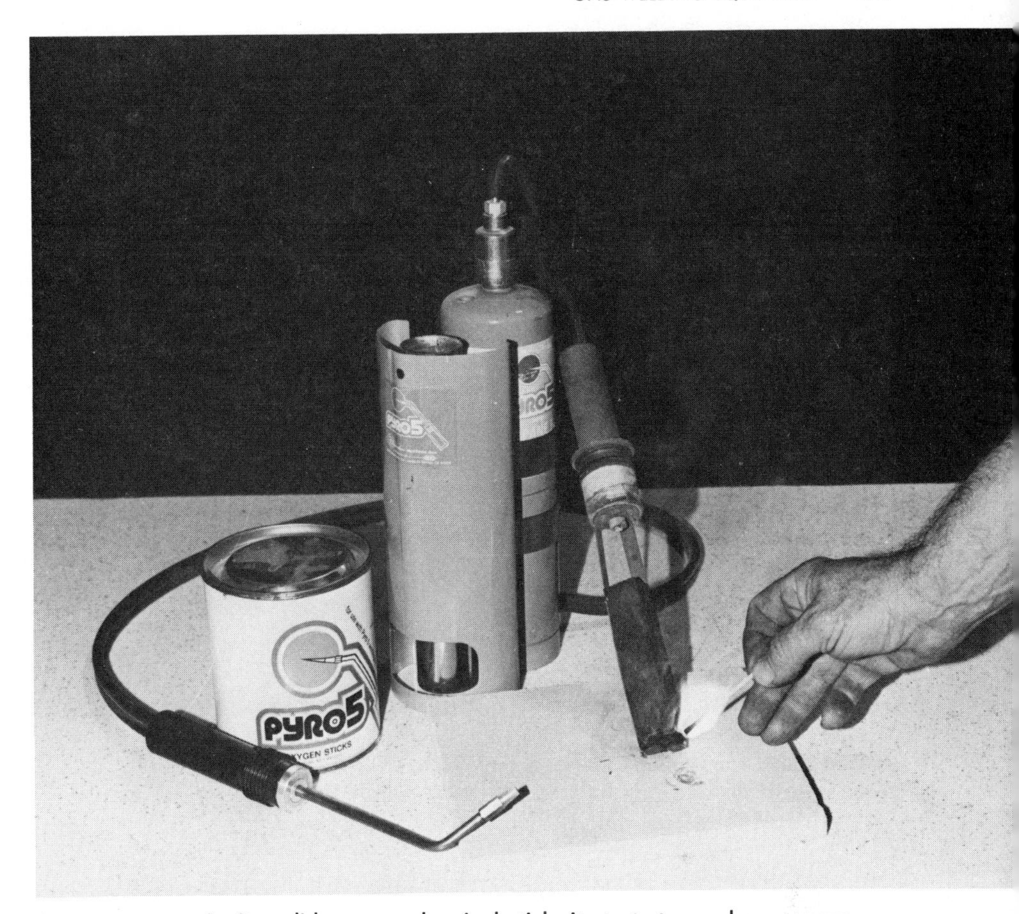

As soon as you ignite solid-oxygen chemical stick, it starts to produce oxygen. Next step is to insert it in container, behind shield attached to Mapp holder, and get to work quickly while oxygen is flowing. There's no way to shut it off short of pulling out stick and cutting away unburned section.

of bicycle frames or garden tiller handles, were difficult or impossible because there wasn't enough concentrated heat to puddle and flow a good fillet. In most cases I either exhausted the rather expensive oxygen tank or used all of the chemical oxygen-producing material before even getting well started. Attempts to repair a cast-iron sector gear from a garden tractor, a routine home-shop job, were not successful either because the heat was not high enough to get the braze metal really flowing or because the result was a superficial joint without strength. To be fair, I repeated the tests with conventional oxyacetylene equipment and did in a minute or two what had been impossible to do with the small MAPP-oxygen torches in ten minutes. It is worth noting that these products have been withdrawn from the catalogues of large mail-order firms.

Pinpoint flame from this tiny butane/oxygen torch is great for either soft soldering or silver soldering on jobs where it's important to have intense heat that's confined to a small area.

Nevertheless, there are trades and hobbies with a real need for a very intense, pinpoint flame even though it may not be hot enough for fusion welding. Jewelry work, delicate repairs, and miniature building often require silver soldering in a very small joint. If that's for you, a tiny torch outfit about the size of a cigarette package and fitted with two gas bottles of the size used for dispensing carbon dioxide from a seltzer bottle is the answer. The fuel gas is butane and the oxidizer is nitrous oxide. It won't fusion-weld steel, but it's wonderfully handy for the work I've mentioned.

15 Preparing to Gas-Weld

Before making a move toward a practice weld, go back to Chapter 5, "Learning to Arc-Weld," and read the general guidelines which apply to all welding. As with arc welding, your attention is focused entirely on the molten puddle of metal at the tip of your torch. But while you're watching the puddle there will be sparks and globules of metal flying about. So you must consider shop fire safety and the safety of those about you. In gas welding you're not concerned with the electric supply, but you are concerned with the oxygen and acetylene supply to your torch. Again, you've got to get everything together in a safe and practical order before you proceed.

SECURING THE TANKS. As soon as you deliver your tanks to the work area, you must secure them to prevent tipping over. Acetylene, as explained, should always be stored upright. More important, it is extremely easy to tip over tanks and seriously damage or break off valves, regulators, and gages. Any of these accidents can result in anything from a slow, unnoticed leak to a gush of combustible gas or oxygen into your shop. If this happens because of a tug on the hoses while welding, you might find it very difficult to escape in the time it takes to remove your goggles, turn off the torch, and assess the situation.

So, even though you have good intentions of eventually building a tank cart to hold your tanks, right now is the time to use a chain, clothesline, duct tape, or whatever to lash them to a basement column, bench leg, vertical pipe, or an eyebolt into the wall or bench.

CONNECTING THE TANKS. Remove the protective caps from your tanks and store them where you can find them when you return for a refill. Usually, only the oxygen cylinder is capped. Now, stand clear of the outlet

Tipping over tanks with gages mounted can be extremely dangerous and cost you heavily for repairs. As soon as you get tanks into your shop, secure them somehow. Best solution is to build a cart.

connection and snap open the handwheel quickly about a quarter-turn. Close it immediately. The resulting burst of gas helps to clear out any dirt in the connection so you'll get a good seal and not damage the regulator fitting. Even after doing this I take a flashlight and look into the tank fitting for trapped debris. I also wipe off the regulator connections with a soft, dry cloth.

Insert the oxygen regulator connection so you feel it seat, then carefully start the threads of the coupling nut with your fingers. Remember, this is a right-hand thread. Turn the coupling down as firmly as possible with your fingers, at the same time moving the regulator slightly by hand. It

Home-built tank cart is stable but easily moved about, even up stairs. Secret of stability is proper placement of axle behind center of gravity and having the wheels just high enough so they clear the floor and rest the tank on small feet. See plan, page 169.

should feel snug and well seated. Give the nut a final solid tightening with an open-end wrench that fits the nut accurately. If you didn't acquire a multiple-opening wrench made for this purpose with your outfit, plan to buy one immediately. Make a final check to be sure the regulator screw is fully backed out.

The same procedure should be used for the acetylene regulator, except, as mentioned earlier, you may need an adapter to match the tank outlet to the regulator. Again, run the coupling nut (left-hand thread) down firmly with your fingers, but this time make an extra check to be certain that one of the gages is not bearing against the tank flange because the connection is too short. Be certain that the regulator is fully backed out.

Checking for leaks. It's only good sense to test for leaks at the regulator-to-tank connections before coupling on the hoses. Mix up a solution of

Special wrench for tank fittings and couplings is better than an adjustable wrench. Couplings must be snug, but don't tighten until you're certain they're properly seated and the nuts turned down by hand. If you need a wrench before this, something's wrong.

After coupling on regulators, open tank valve and use a soap solution to test for leaks. It's a good idea to do the same with the tank valves themselves as soon as you bring the fresh tanks home. Wash away soap residue with clear water before using equipment.

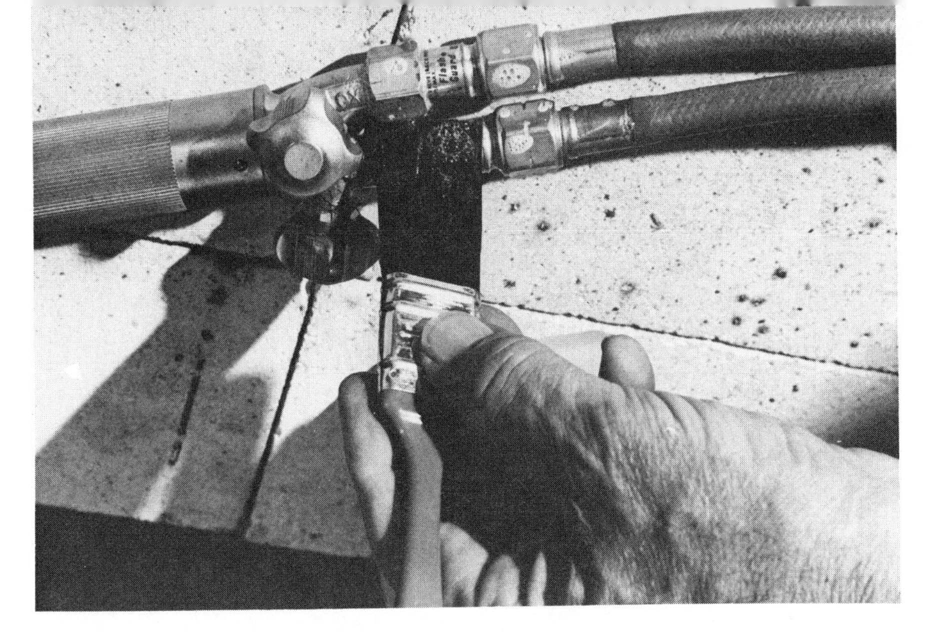

Repeat soap check on torch connections and around torch valves and tip after opening regulators to feed gas pressure to these parts. If you can't stop leaks with reasonable tightening, get a repair from a competent service agency. Don't try home repairs.

Ivory soap or dishwashing liquid and water. Slowly open the oxygen handwheel until full tank pressure registers on the gage. Continue to open the handwheel until you feel it seat at the top. Carefully brush the soap solution around the tank connection and around the stem of the handwheel and at the regulator connection to the tank valve. If you locate a leak at the regulator connection, flush away the soap and wipe dry. Close the handwheel, open the regulator adjusting valve slightly to bleed the trapped oxygen from the regulator, close the regulator valve, and remove the regulator to inspect the seat for anything which might cause leakage. In all these tests a good flashlight will help in spotting bubbles and bits of foreign material. Repeat the installation procedure and test again. Unless there is a defect in the coupling you should be able to obtain a gastight seal.

If you find a leak around the handwheel stem, return the tank for exchange. Never try to repair a tank at home. Repeat the above tests with the acetylene tank. After being certain that there are no leaks, flush away the soap residue with clear water.

CONNECTING THE TORCH. Use only red and green hose made for gas welding equipment, and be certain that there is no visual damage to the end fittings, their threads, or the threads on the torch or regulators. Never use oils or sealants on these threads. Always run the couplings down fully with your fingers, and follow with a wrench. Never force a coupling which doesn't turn on freely. Again, don't forget the right-hand and left-hand threads for oxygen and acetylene respectively.

When both hoses are securely coupled to the regulators and the torch, turn in each pressure regulator to read about 5 psi on the outlet gage with the valve of the torch cracked open. Now close the torch valves and repeat the soap test at both ends of the hose connections. If there are no leaks here, dip the tip of the torch in the soap with both torch valves closed. Any bubbles at the torch tip indicate a defective torch valve which must be repaired.

Finally, test the packing around the torch valve stems with soap. A minor leak at this point can usually be stopped by gentle snugging of the packing nut. If this fails, or if wiggling the valve knob produces bubbles, the torch must be repaired.

Caution: The soap residue may act much like oil in the presence of oxygen. Wash away the soap with clean water and dry the equipment.

16 Learning to Gas-Weld

Since oxyacetylene welding is accomplished with a mixture of two gases, it is essential that the mixture be very carefully adjusted. The first, coarse adjustment is to set the regulators to the proper delivery pressure. When you buy your welding equipment you will get an instruction manual advising the proper pressure settings and tip sizes for different metal thicknesses. This varies somewhat for different outfits, but the following table will serve as a rule of thumb for most home-shop equipment. Check the information in the table against your instruction book. Rather than hunt this up in the book each time, make your own table and tape it to your cart or bench. Suit the tip sizes to those identifying numbers or letters used with your torch. Other makers have different tip size designations.

Metal Thickness (in.)	Tip Size	Rod Size (in.)	Oxygen Pressure (psi)	Acetylene Pressure (psi)
3/64	1	1/16	1	1
1/16	3	1/16	3	3
3/32	5	3/32	5	5
1/8	5	1/8	5	5
3/16	7	5/32	7	7
1/4	9	3/16	9	9

ADJUSTING THE REGULATORS. For a start, after opening the oxygen valve slowly and then to full position with the regulator fully backed out, open the oxygen valve on the torch about one turn. Now turn in the regulator control and adjust the outlet pressure to 3 psi. Truthfully, you may find this recommendation somewhat humorous, since most oxygen delivery gages are marked off in 5-psi or greater divisions. The best you can do is guess. Close the torch oxygen valve.

Open the acetylene tank valve slowly and to no more than one turn. Leave the wrench in place ready for a quick shutdown at all times. In most work you'll probably find that one turn open is all you need. Open the acetylene valve on the torch one turn and adjust the acetylene regulator control to 3 psi on the delivery gage. This gage is usually marked in smaller divisions. Close the acetylene valve on the torch.

ADJUSTING THE FLAME. You are now ready to make the second, critical adjustment of the actual flame. Check again for inflammable materials in the vicinity of the torch. Don't forget to look overhead. Unlike an arc, your initial flare when you ignite the acetylene will go toward the ceiling. Not only do you want to avoid starting a fire if you have wood or other material stored above, but you also must expect some very heavy black smoke which will coat anything it touches.

Start right now to form the habit of never pointing the torch tip toward your face or person. As a final check, look to see where your torch hoses are. It's surprisingly easy to have them coiled or draped on the bench or elsewhere where you might hit them with the flame. Try to form the habit of bringing the hoses up over your shoulder or at least from a location behind you and your work.

Have your spark lighter ready, open the acetylene valve about half a turn, and direct the tip into the pan of the lighter. You'll get quick ignition and probably a very smoky, orange flame. Open the acetylene valve slightly until the flame cleans up somewhat. If your flame now appears to have jumped away from the tip of the torch about ⅛″ or so, you probably have too high a pressure setting on the regulator. Back off just a trifle. You want a fairly clean flame but not excessive pressure.

Now start opening the oxygen valve on the torch slowly. The flame will immediately change and go from yellow to blue. From here on your flame adjustment calls for a very careful observation of the inner flame cone right at the tip of the torch.

The desired flame for welding is called a "neutral" flame. It has neither too much acetylene nor too much oxygen. As you continue to open the oxygen valve a gentle twitch at a time you'll see the whitish inner cone become more distinct and sharp-edged until finally there is only a small feather around it. You should have your welding goggles on when making this adjustment, both for safety and because they make it easier to distinguish the flame.

Your final adjustment must be very precise and just enough to remove the feather caused by excess acetylene. The neutral flame has a distinct cone, without feather, and with a slightly rounded tip. If you err, however,

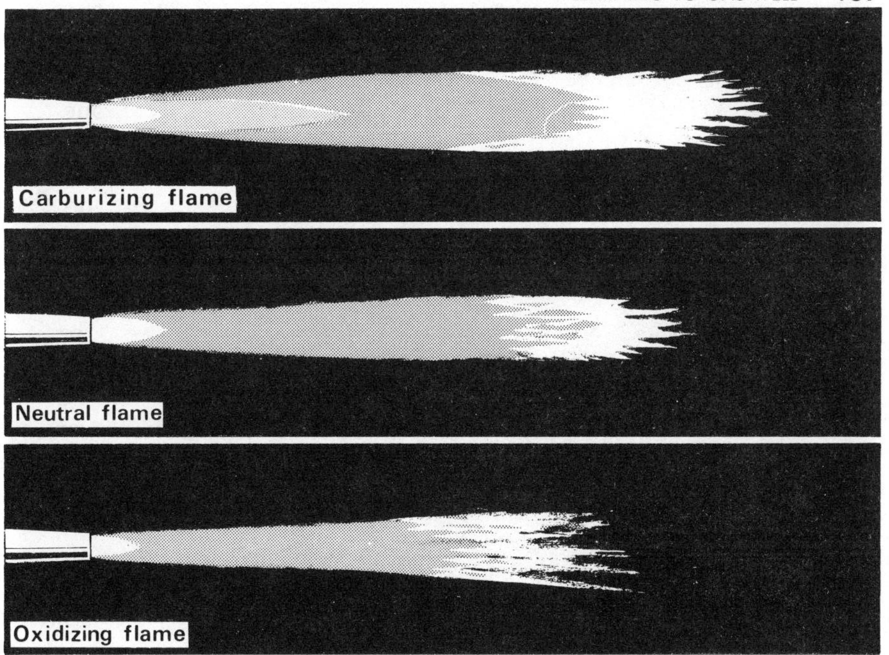

In addition to flaring black and orange of pure acetylene flame, you'll soon come to recognize long, inner feather of an excess actylene flame, called a carburizing flame; the sharp, pointed inner cone of an oxidizing flame with excess oxygen; and a neutral flame without feather or point, which is usually best for welding. Check your flame as you work; drifting out of adjustment is common.

it's better to err very slightly on the side of excess acetylene. Some call this a carburizing flame.

To learn more about your welding flame, continue to open the oxygen valve in slow, small steps. You'll see the inner cone assume a more pointed conical or needle shape, and the sound will change to a hiss. This is an oxidizing flame and means too much oxygen is present. If you weld with such a flame you'll soon see little sparks flying like those from a Fourth of July sparkler. You're actually oxidizing or burning your work metal, and the weld will be weak and have a poor appearance. Such welds are typical of those attempted with MAPP/oxygen. On the other hand, a very slight oxidizing flame is suggested for braze welding where you do not elevate the steel or iron to molten temperature.

RUNNING YOUR FIRST BEAD. As discussed with arc welding, you'll need plenty of practice metal. Since gas welding is better suited to light stock, you should practice on 1/16" or 3/32" steel. Heavier material takes more heat and costs you more for gas.

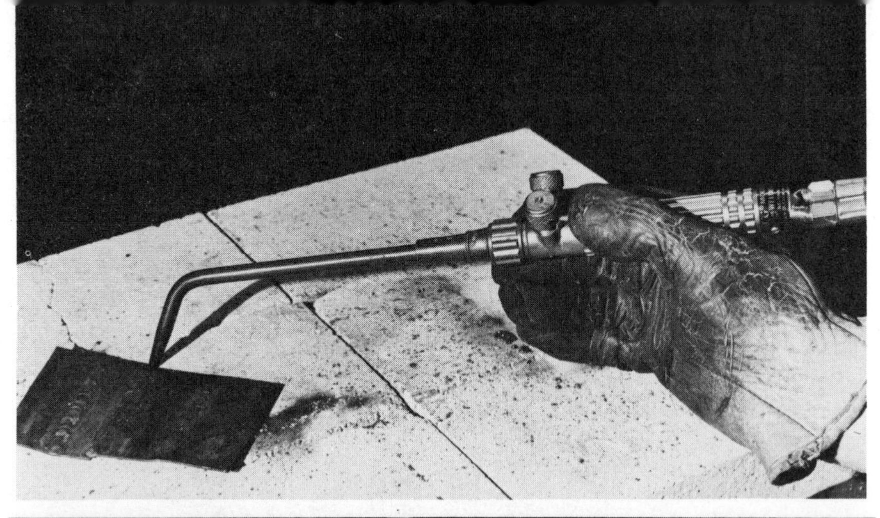

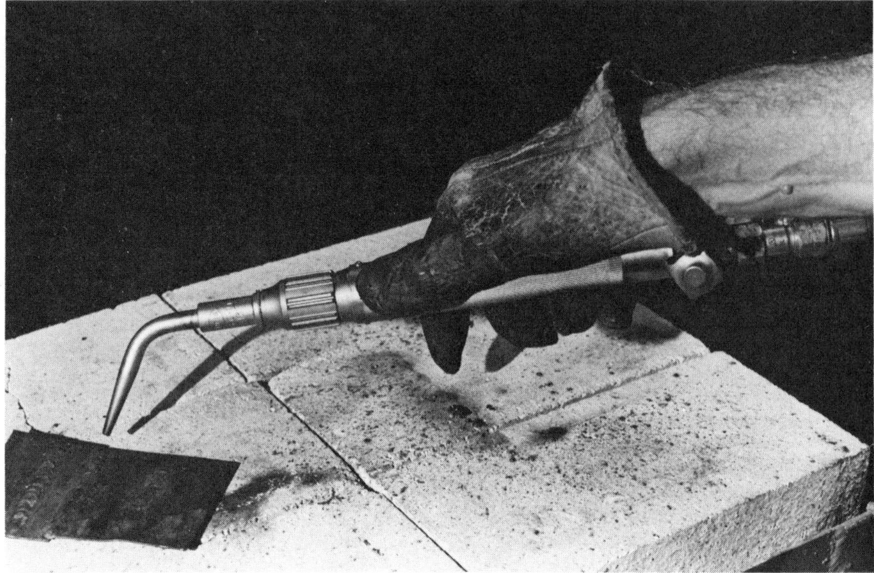

Whichever method of holding torch feels right to you is the one to use. Many welders hold small torches pencil-fashion, top; heavier torches seem to handle better when held in full hand. There's no rule, however, saying that one is right, the other wrong.

Place the metal on the welding bench or firebrick in front of you so you can assume a comfortable seated position. Being seated helps in developing a delicate control of the torch. The work should be at or slightly below elbow height. For the present, only one piece of metal is needed.

If you've read the section on arc welding—and I emphasize that you should—most of the techniques and practices will be familiar to you, with obvious differences, of course. You know that the electric arc leaves a bead of filler rod. The gas torch leaves no filler rod and you must feed in filler from a rod held in the free hand. Note that with gas welding you add filler at your own rate, amount, and discretion. That's a big difference and it makes gas welding even more of an art. Another important point is that right-handed arc weldors normally move the puddle from left to right. With gas welding you start at the right and progress toward the left.

Before learning to add filler rod you must first learn to develop and control a puddle so as to get good penetration without burning through. Your first attempts should be to run straight beads without filler along a piece of 1/16″ steel marked off in horizontal guidelines. Before you start, try holding the torch in two different ways. A light "aircraft" torch such as you might have for small hobby work may feel more comfortable to you when held much like a pencil. Most users will grip heavier torches full-hand like a hammer or a fishing rod. Try working both ways after you've gotten the touch.

After lighting and adjusting your torch, start at the right side, if you're right-handed, and bring the inner cone of the flame to a point about 1/8″ above the metal. Cant the torch at about 45° so the tip points to the left, your direction of movement. Watch the spot directly under the tip and you'll see it glow, start to appear wet, and then develop a puddle. A very limited circular motion of the torch tip will help develop the puddle diameter. If you continue to hold the puddle without moving the torch slightly to your left, you'll quickly burn through the metal. Continue the slight motion of the torch tip and try to overlap fresh metal by about 1/16″ with each rotation. This will move the puddle along from right to left. This is the common "forehand" method of gas welding. Sometimes weldors move the torch in the opposite direction, to the right, but this is only

Developing smooth, even torch movement requires practice resembling old-fashioned school exercises in penmanship. Instead of pen, you're moving the torch flame cone and puddle along in a series of slightly overlapping circles. For practice, no joint is needed, but you can also butt two pieces to check your penetration.

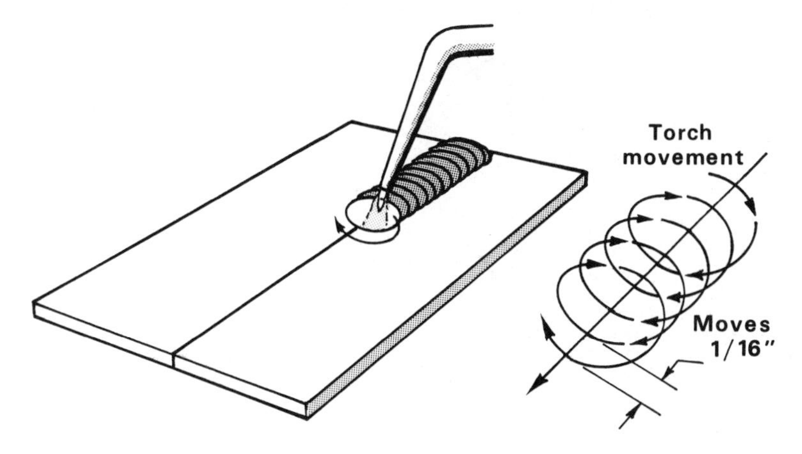

for special cases. The important thing right now is to learn to move the torch and bead along, forming a neat, even bead with good penetration showing on the back side, and without burning through.

Beginner's problems. Don't wear good clothes when learning to gas-weld. Wool pants which resist burns are recommended. In the beginning you'll be startled by pops and flying metal. This is caused by allowing too much flame or the torch tip to contact the molten puddle. Sometimes this will actually extinguish the torch, but it usually restarts from the heated metal. If it doesn't, shut it off and repeat the lighting and adjusting sequence.

Burn-throughs will probably be your second most common problem. If it seems impossible to avoid them, try reducing the oxygen a bit to produce an acetylene feather and then cut back the acetylene to a neutral flame again. Eventually you'll find a heat delivery rate that gives you better control even though your puddle moves more slowly.

On the other hand, if it takes too long to get a puddle and your bead looks lumpy and has poor penetration, try opening the acetylene valve on the torch and then increasing the oxygen to get a slightly larger flame.

You'll find, with practice, that this ability to adjust the flame to your exact needs is one of the advantages of gas welding. You'll also find that a very slight upward movement of the flame away from the puddle slows down the melting action and enables you to regain control of a bead that wants to melt through or spread too wide.

Other troubles may arise because you're not holding the angle of the torch constant. The 45° cant directs some heat ahead of the puddle and preheats the metal. If you accidentally allow the torch to drift into a vertical position, burning through is the usual result. Too flat an angle produces skipping and uneven penetration. Be your own most critical inspector and try to make each bead better than the last. When you're satisfied that you can hold a puddle and move along a line you are ready to start practicing with filler rod.

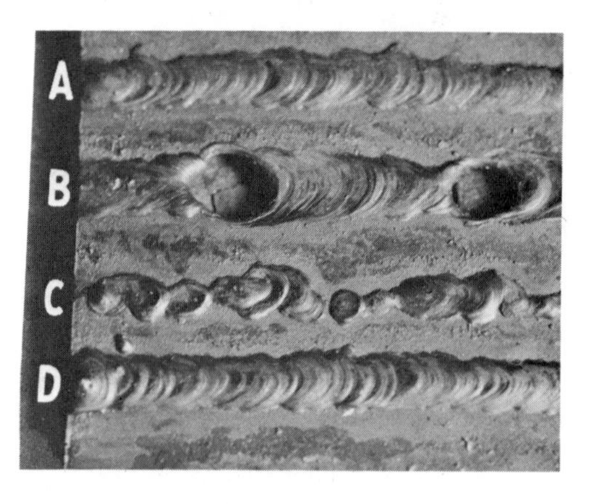

Practice welds show fairly good control of width, height, and ripples in examples A and D. Too much heat was used in B, with result that bead spread too wide and eventually burned through. Too little heat and failure to develop and maintain puddle is seen at C. Rod was merely melted off and dribbled onto cold workpiece.

Careful inspection and thoughtful analysis of your practice work is the only path to developing gas, or electric, welding skill. Welding table and tank cart seen here make good first projects. Plans for both are on pages 168 and 169.

THE FILLER ROD. You may have noticed that an entire chapter of this book was devoted to arc welding electrodes and that even so I stated that this was just a superficial look at the subject. Gas welding is different. You'll almost always be using a standard mild steel rod. It will have no coating. The size will depend upon the size of the job, as shown in the earlier table. For practicing, I suggest a 1/16" or 3/32" rod at the largest. There are other types of welding rod material used in gas welding—aluminum, cast-iron, and brazing rod, and some special rods for alloy steels. For the present, however, there's nothing special to learn about gas welding rod.

Handling the filler. The delicate niceties of gas welding only become apparent when you learn to add filler rod to the bead. One hand must control the torch, its angle, and its movements. The other must hold the rod and be dedicated to definite, specific motions which are "out of phase" with the torch. This is because the rod tip is dipped into the molten puddle to melt a small amount from it as filler. The effect is to cool the puddle. At that moment the flame cone has been raised slightly. After dipping, the rod is lifted a trifle out of the flame path and the flame cone is again directed at, and ahead of, the puddle. This reheats the puddle. This alternate slight movement of dipping first the torch and then the rod produces a smoothly rippled weld in the hands of a skilled gas weldor.

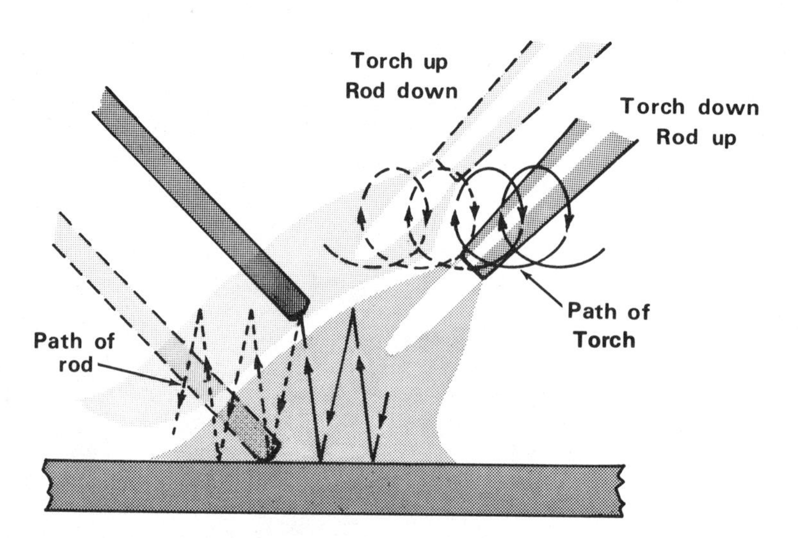

Two different techniques of mutual, but out-of-phase, movement of filler rod and torch may be a matter of choice and what works best for you. One method is up-and-down movement of first torch and then rod, allowing a bit of rod to be melted off in puddle on each dip. The torch redips to heat puddle. Another method achieves same result by arcing torch back and forth and dipping rod while flame cone is at sides.

Since it is not really a natural motion, you must learn the proper coordination, and your first results may be rather unattractive. Moreover, there are other movements which many weldors prefer. These are basically slight circular movements of the torch tip with the rod being dipped as the tip and the flame cone are brought out to the edge of the circle on each side. The rod is withdrawn as the cone again comes back to the center of the puddle path. Try both motions when you practice. My experience is that the circular motion is a little better when you need a wider, flatter bead such as a lap weld. The dip and lift seems to work down into the corner of a sharp fillet weld best.

Practice with filler. A flat piece of 1⁄16″ stock should be marked off and placed comfortably before you. A piece 3″ or 4″ square is adequate. Use 1⁄16″ rod. To reduce the instability of the 36″ rod, cut it in half. Some weldors prefer to bend the rod at a shallow angle about 6″ back from the end so the hand can be out of the projected heat from the torch. I've never found this helpful, but I do recommend crimping over the upper end of the rod to identify it so you won't pick up the hot end accidentally.

Now run a series of beads using the movement or movements described above. Most important, avoid the temptation to melt the rod off above the puddle and drip it in. The only exception to melting the rod in the puddle is when you are filling a hole or gap and want to deposit extra metal which will later be melted to fuse with the work. Unlike the arc rod, which carries its own shielding against oxidation, your gas rod is unprotected. While welding, keep it within the shielding effect of the outer flame envelope. This also preheats the rod so it melts readily.

Perhaps by now you have discovered that steel passes through a gradual transition in going from solid to fluid. In the first stages it's soft; later, somewhat plastic; and still later, almost but not quite free-running. In the last state there is a point where it is just plastic enough to fuse and accept filler rod but is still not watery. This is not easy to recognize or take advantage of in arc welding but it is one of the advantages of gas welding. When you have developed the ability to sense the different stages of fluidity you'll be able to take advantage of them in handling difficult vertical and overhead welding. Sometimes when you're trying to determine the plasticity of a puddle, but more often when you have no excuse at all, the rod will stick solidly. Don't try to break it loose. Instead, flick the torch flame cone on the point of sticking for a second and it will melt free.

STANDARD WELD JOINTS. When you've become fairly proficient at running unfilled and filled beads, go back and read the chapters on arc welding of joints, fit-ups, clamping, and the rest. There are no real differences, at least in home-shop welding, between butt welds, edge welds, lap welds, and filleted tee welds completed with arc or gas. Practice the same

Expect to use up lots of scrap metal, produce some ugly beads, burn holes, and be frustrated when you start to add filler rod to your gas welds. It probably takes longer to learn to weld well with gas than with arc equipment.

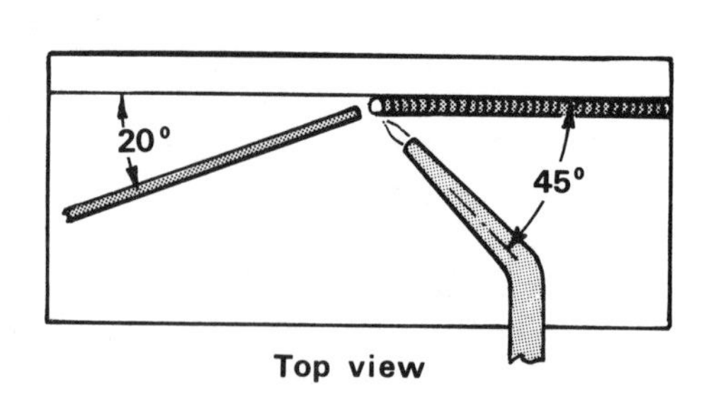

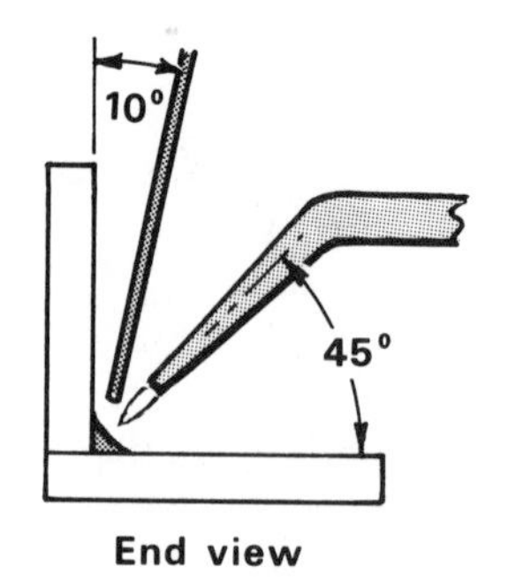

Gravity is always a factor when welding a fillet with one face vertical. The molten metal wants to run down. Try holding the flame cone directly into the corner with the torch angle about even between vertical and horizontal surfaces. Add filler with rod held at a shallow angle and at the top of the fillet so natural flow is downward.

sequence of joints, watch the effects of expansion and contraction, study the best way to tack-weld, and examine your gas welds for full penetration to the bottom of the joint. With fillet joints you will encounter many of the same problems as in arc welding. You will learn that you must get the flame cone down directly into the corner of the joint, that you can gain by introducing your filler rod a trifle high on the vertical face, and that by directing the heat you can control metal deposition from the top to the bottom edge of the fillet.

In some respects all of these joints are a little easier with gas than with arc, especially for the beginner. Since you can see your work and approach it delicately with the flame, there is less chance of knocking your set-up out of line. In fact, on many projects, even though you want to make the final welds with arc, it's often best to tack with gas first. And once an arc is struck you must proceed at a fairly fixed pace to avoid piling up filler rod and overheating. With gas there's nothing wrong with drawing back the torch a bit, keeping the weld in the cooler outer envelope, and taking time to contemplate how it's going and what you might be doing better.

EXPERIMENT. As with arc welding, there are many suggestions by different experts on exactly how to weld certain joints. My suggestion is that you follow along exactly the sequence of practice joints recommended for the beginning arc weldor. Try both the dipping and the rotating techniques for the rod tip and the torch. Gas welding is very much an art and a skill. Just as artists and wood carvers use different strokes and cuts, you can teach yourself different torch and rod movements. If, in the end, you've produced a sound bead or fillet with good fusion, don't fret about whether you did it exactly by the book.

One other movement you should try on heavier work such as steel plate.

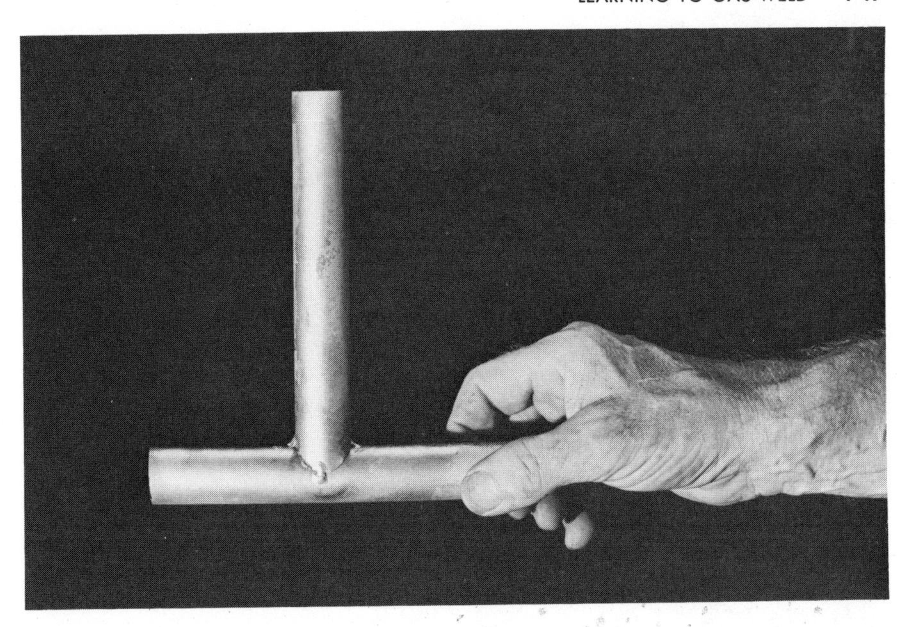

Fishmouth weld on conduit or tubing is excellent practice. After tacking as shown, try welding the fillet with the practice piece in many different positions—flat, inverted, and at angles. All are good practice for orienting flame and filler and judging flow of molten metal.

Use a larger tip and more pressure. Because the plate is heavier and absorbs more heat you will need to maintain almost continuous preheat ahead of the torch. Hold the torch at about 60°, a bit nearer vertical than you've been doing. Weave the flame cone back and forth across the weld in a series of small arcs. You'll need a larger puddle, perhaps ⅜″. Once the puddle is established, insert the rod tip in it and leave it there. Move the rod back and forth in the puddle and you'll feel it melt down. The rod should move opposite to the torch in a timed sequence. Let the little arc the torch tip makes detour slightly around the rod tip as you move each from side to side.

The procedure above is very satisfactory for heavy gas welding except that it tends to use up gas and oxygen rather rapidly. Gas welding of thick plate or pipe is not really economical, although it may be necessary at times. If you find you're getting into this type of welding, it's time to go back and reread the section on the reasons for arc welding.

Saving rod. Unlike arc electrodes, there's no reason to discard a stub of gas welding rod. And there's not much point in working it down so short that you burn your fingers. When the stub gets uncomfortably short, stick it in the weld so it projects firmly from the metal. Take a fresh piece and hold it in position against the outer end of the stub. Flick the flame cone

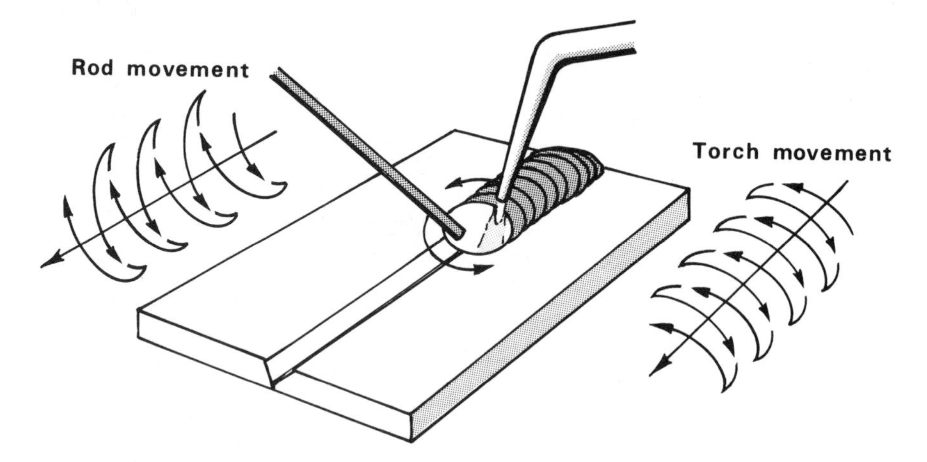

For heavy welding with large tip, try leaving the rod tip in the puddle and moving it back and forth. At the same time, hold the torch a bit more than usual toward vertical and move the flame cone in a series of arcs.

onto the joint for a second or two to weld the new to the old. Let it cool a little, melt loose the end, and go on welding.

Now and then you hear home-shop weldors claim that they never buy rod. They use old coat hangers and the like, including scrap fence wire. Such frugality is self-defeating if you want to learn to weld. Very often such scrap wire will behave oddly under the torch with sparks and sputterings which indicate that you are doing something wrong. You are, of course; you're using the wrong rod.

SHUTTING DOWN. Gas welding equipment can't be just switched off and forgotten. Even though you shut off the torch valves and stop the flame, pressure remains in the hoses and regulators and leakage can occur. Over a period of time the leakage of acetylene and oxygen into your shop could cause a disastrous explosion. Your best protection is a rigorous, habitual shutdown procedure. In home-shop welding there are always interruptions such as phone calls, visitors showing up for a minute and staying for hours, and all the little things that can make you forget that you hastily shut off the torch and laid it down. That's why developing the habit of full shutdown is so important for your own safety.

When shutting off the torch, close the acetylene valve first and then the oxygen valve. Some welding instructions suggest that if you are only going to stop for a few minutes you only back out the regulator handles and open the torch valves to bleed out the regulators and hoses. But this leaves full tank pressure on both regulators. In the words of one old-time weldor, "You're hanging your life on a couple of gages." This is poor practice, not only because of leaks, but also because it's easy to knock the tanks over accidentally, sometimes just in moving projects or materials around the shop. If a regulator is cracked loose with full tank pressure on it, the results can be dramatic and all bad!

After shutting off the torch, the only safe procedure is to close both the oxygen and acetylene tank valves promptly and firmly. Then open both torch valves and the entire system will bleed down to zero. Back out both regulators to full off. Close the torch valves.

If you have kids with busy hands around your home shop, remove the regulator screws completely and lock them away. It's a bother but it could prevent a tragedy. If you have lots of dust in the shop it's a good idea to slip plastic bags over the regulators when the handles are removed.

And when you remove the regulators for storage, or until you get around to picking up some fresh tanks, you can help them be ready to perform in the future by plugging all the openings with little cup-shaped plastic plugs and sealing them in plastic bags. Before doing this, however, regulator manufacturers suggest turning in each regulator handle just enough to move the valve off its seat. This preserves the valve seat and helps to avoid future sticking.

Be safe, not sorry. Don't take the chance that a curious youngster won't turn a handle to see what happens. The smaller they are, the more they love to do it. Remove regulator handles and lock them away if you've got kids or grandkids.

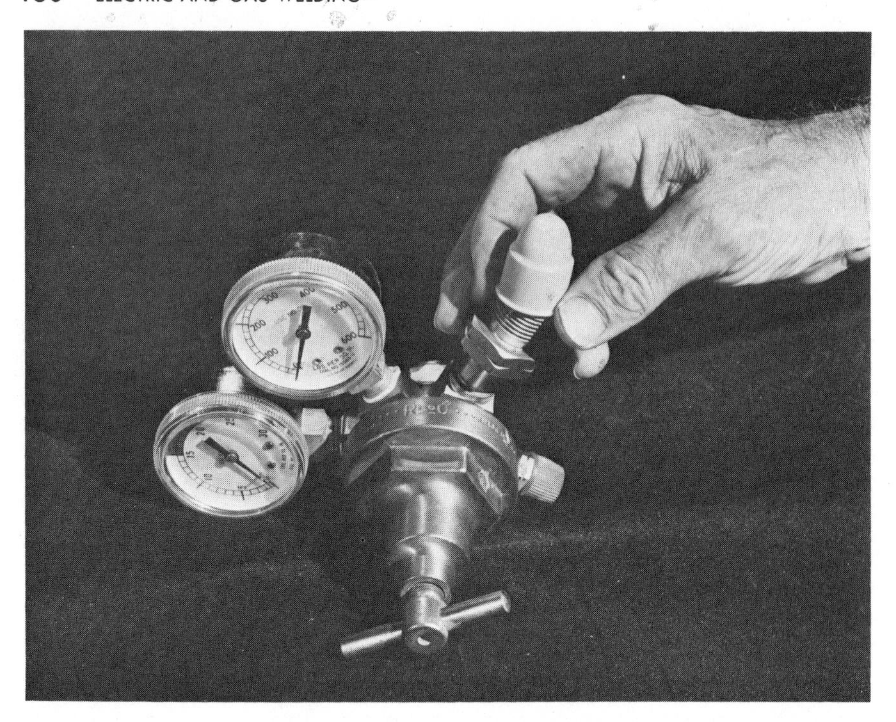

Save the plastic caps that come with your regulators, or buy new ones, to seal these units while stored. Next to rough handling and opening the tank valves too fast, dirt and dust are the greatest enemies of these precision devices.

Sometimes you'll see weldors put masking tape or duct tape over regulator openings and hose and torch connections. This is not good practice because residual adhesives can react like oil in the presence of oxygen. Moreover, the adhesives can leave gum which interferes with finger-tightening and also picks up grit and dirt which can interfere with the important surface-to-surface seals.

17 Brazing, Soldering, and Aluminum Work with Oxyacetylene

Nearly all of the instructions offered on braze welding with the carbon-arc torch apply equally to doing the same job with oxyacetylene. Gas torch brazing tends to be a little neater than carbon-arc brazing because it's easier to pinpoint the flame. This is especially true of very small joints, because you can use a small tip and confine the heat to a localized area. For brazing, the torch is usually adjusted for a slightly oxidizing flame.

The cost of gas can be a factor on some jobs, such as casting repairs where the mass of the metal may require a large amount of heat for an extended time compared to what is needed for light sheet metal. Here, carbon-arc brazing will cost less than gas. Whichever heat source is used, it's good sense to consider one of the new high-output MAPP or propane torches for preheating. Long preheating with oxyacetylene is expensive, and when done with a carbon-arc torch may overheat your welder.

FLUX. Since braze welding is essentially a soldering process, even though the final joint may be stronger than the work, you must have clean surfaces and use some sort of flux. There are many pre-fluxed specialty rods on the market and many special alloys. For average home-shop work, however, I recommend conventional unfluxed brazing rod and a can of borax-base flux such as Brazo.

Clean the work area to a bright surface with a file or abrasive and sprinkle a coating of flux onto and into the joint. Since the gas torch tends to blow away the powdered flux, you might try heating the metal first just enough to make the flux stick.

Your next, and often repeated, step is to heat the rod tip in the torch flame and dip it in the flux can to pick up a coating of flux. Return the torch flame to the work area and keep the tip of the rod in the outer flame to warm it. When you see the flux on the work melt and spread out, melt a

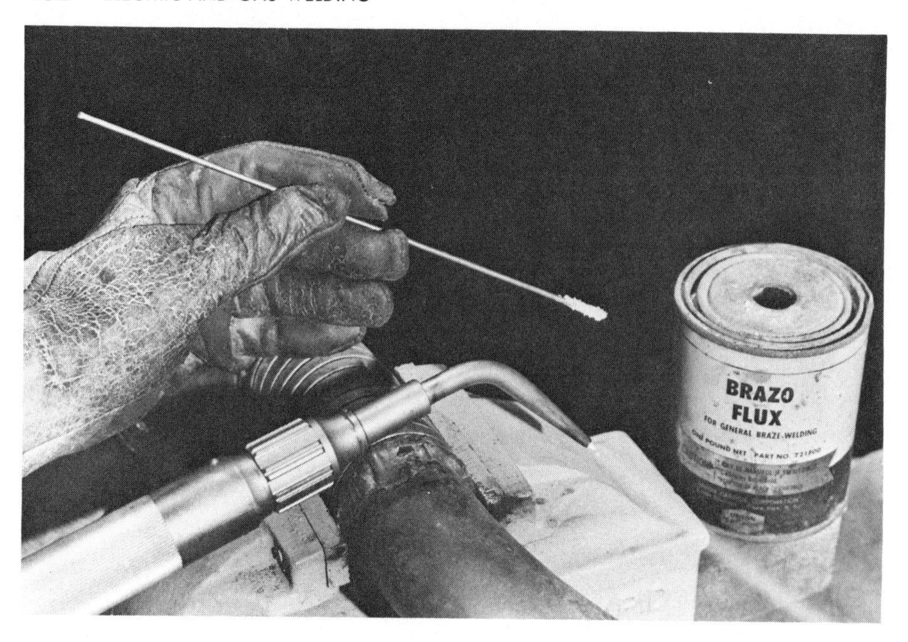

Heat brazing rod for a few seconds and dip it in flux can to get a coating like this on tip. Repeat as you use up rod. Brazing is ideal way to join dissimilar materials such as flexible exhaust tubing and thin-wall pipe shown here.

Often different parts of same job call for either welding or brazing. On this theme table, heavier steel strips were ideal for gas welding, but delicate points on stars would have burned away before puddle could develop in steel strips. By heating heavier metal to point where flux just melted and then flicking torch onto star points momentarily, a small spot of braze metal could be deposited without burning or distorting the stars.

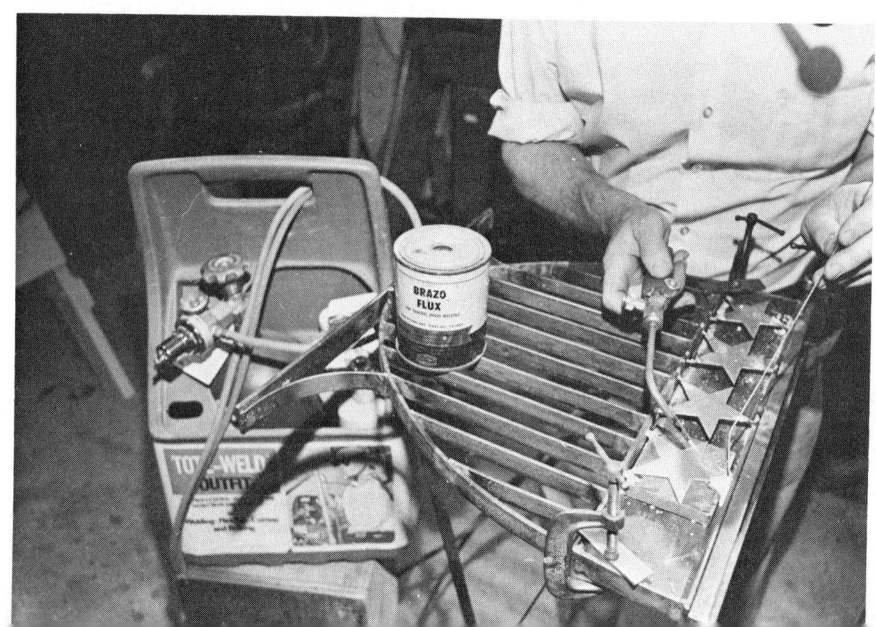

drop or two of metal from the rod. If the temperature is right the rod metal will stay fluid and spread out smoothly. If the rod metal doesn't quite flow, continue to add heat to the area until it does. Once you've reached this required heat, try not to add any more than is needed to sustain the temperature.

Excessive heat is a common problem in brazing. This tends to boil out the zinc in the rod alloy, which is about 60% copper and 40% zinc. The result is bubbling and the formation of droplets rather than a smoothly adhering deposit. If the metal is too cold you will get a lumpy deposit marked by edges which do not feather out into the work metal surface.

BRAZING TECHNIQUE. Some lightly brazed joints are pretty much a matter of a single quick pass, but others require a substantial buildup of metal. The first step is tinning, or getting the work surface coated with smoothly adhering braze metal. Only after a good job of tinning can you proceed to build up metal. To accomplish this buildup, try starting with the torch at a fairly shallow angle and depositing a ridge or dam of metal. This will act as a support onto which you can build. You then add onto this by working with the torch and rod in small arcs.

It may be that after building up all along or around a joint you note areas where the metal has bubbled or didn't flow and adhere. The usual cause is poor initial tinning. No amount of further heating will plaster it over. Grind or file out the bad area to clean metal, re-tin, and merge new metal into the area as necessary. If the deposit is merely lumpy and rough but with good adhesion, there is nothing that says you can't reheat and smooth it out. The usual method is to heat the metal to form a puddle and then work the puddle along, letting the metal smooth out as you go. Remember, the braze metal will follow heat and flow toward the hottest spot.

REPAIRING CAST IRON. Everything discussed under arc welding and the carbon-arc torch on repairing cracks or breaks in cast iron applies in much the same fashion to welding or brazing with gas. A skilled operator, using cast-iron filler rod and a special flux different from brazing flux, can fusion-weld cast gray iron with a gas torch. If you want to do this in the home shop, practice on something of no value before trying a serious repair.

You'll need a means of heating the entire casting evenly and to a dull-red heat. And you'll need some way to cool the finished job very slowly and over a full day or more. If you visit the salvage shop of a company making large castings such as engine blocks and machine bases, you'll see complex tents and ovens of asbestos and firebrick which permit the castings to be preheated slowly by large propane torches and stoves before welding. These operators are very skilled at blending heated areas into cooler areas and have learned about any given casting over a series of successes and failures. You cannot duplicate this at home.

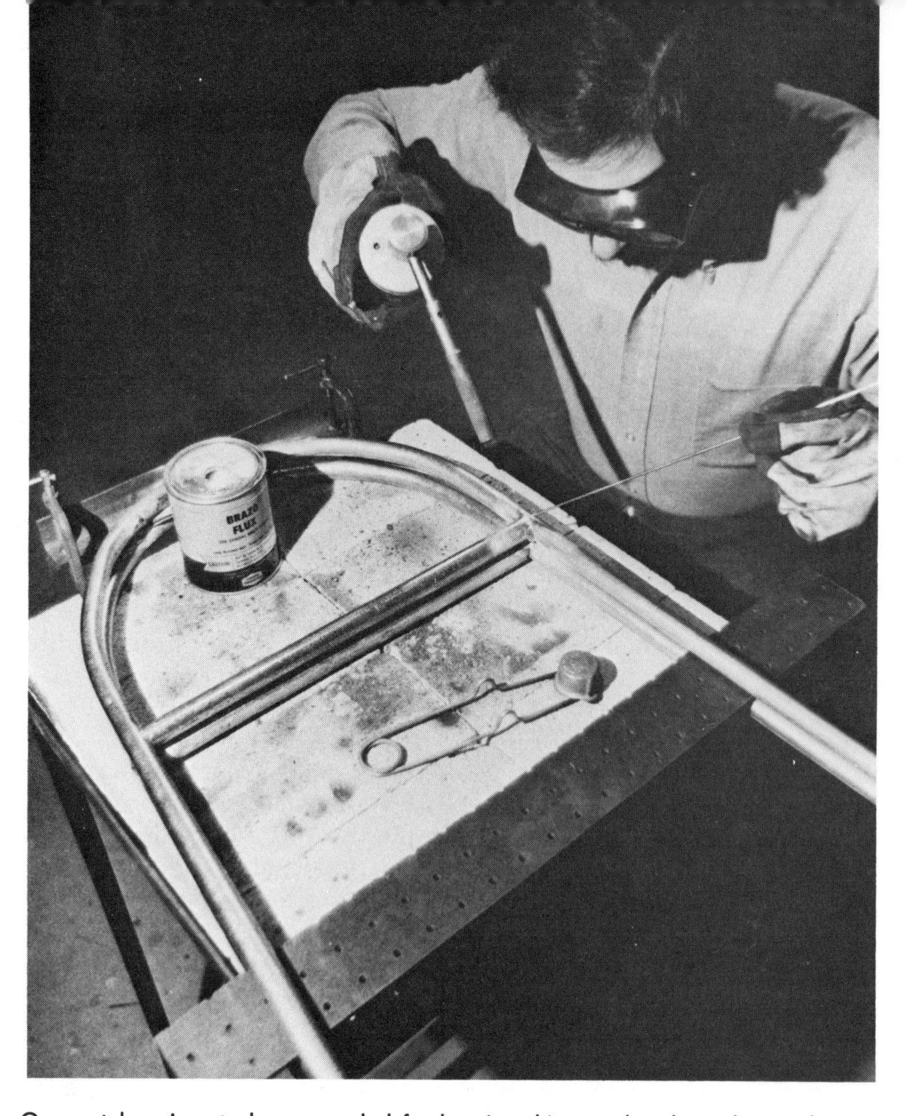

Oxyacetylene is not always needed for brazing thin metal such as the conduit used in this sewing-table end. MAPP/air torch provides adequate heat, and spring of metal keeps work in place. This type of torch does not pinpoint flame, so it's hard to tack and proceed without melting tacks. Entire joint becomes fluid and clamping is needed.

Repairing cast iron is somewhat easier with conventional brazing rod and flux. Veeing and cleaning the fracture, as well as drilling at each end of a crack, still should be followed by preheating. The difference is that since you're not going to fusion temperature, the preheat can be held to a lower level. It's best not to attempt to deposit a full bead that fills the vee groove in a single pass. Start by tinning, being sure that the tinning action gets right to the bottom of the vee and well up each side. Next, run in a fairly light bead. After slight cooling, repeat the cleaning and tinning to establish good adhesion on the sides of the vee. Add another fairly light

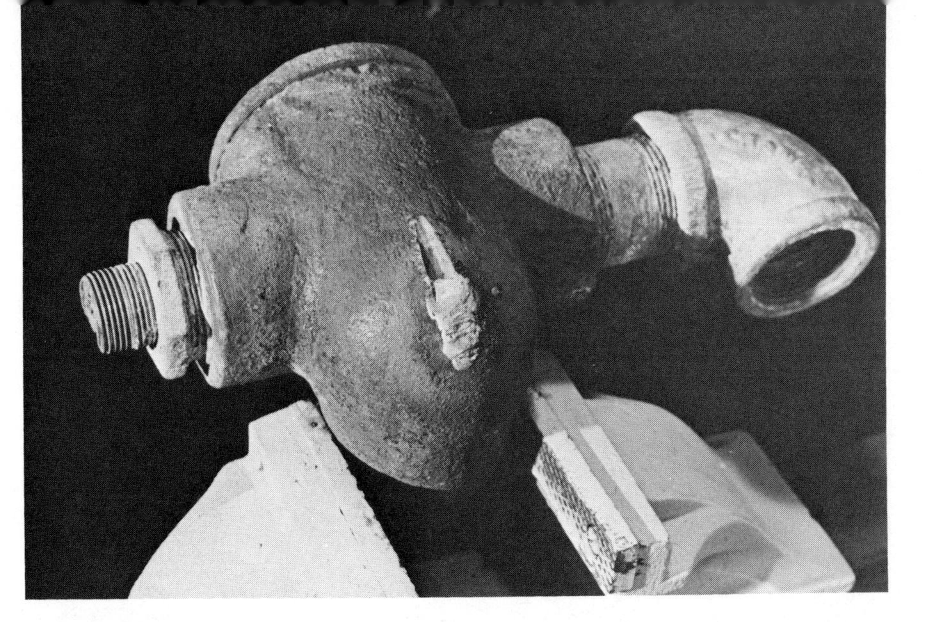

Hairline freezing crack in cast-iron plumbing fixture is shown partially brazed. After veeing out crack area, bottom and sides were tinned. Metal was then built up, more than needed in this case. There is no reason to build excess brazing material outside of work area.

bead. Each bead must fuse thoroughly into the underlayer. Use three or more passes to fill the bead but do not run randomly over the surface surrounding the repair. This step-by-step procedure not only helps to assure good adhesion and strength but also avoids using the amount of heat required for a single pass. Nevertheless, very slow cooling or even post-braze heating which is gradually tapered off is still recommended.

SOLDERING. A good bit of confusion surrounds the terms "brazing," "braze welding," "soldering," and "silver soldering." Technically, braze welding is accomplished below the melting temperature of the work metal but with a filler rod which melts above 800° F. The metal is generally deposited in the joint much as in fusion welding. Brazing uses the same type of filler rod but instead of depositing metal directly it is allowed to flow into a snugly fitted joint by capillary action. An example would be a copper fitting on the end of a copper tube. The fitting goes on with a snug slip fit, but a very small space still exists between the inner diameter of the fitting and the outer diameter of the tube. When the metal is heated to the proper temperature the braze metal will flow by capillary action into and around the inner surfaces. This is a form of soldering, but more often soldering refers to the use of even lower-temperature alloys. The main thing to remember is that soldering involves a joining material that melts below the melting temperature of the work metal, that must adhere to the work metal, and that has the ability to flow into and fill small openings and gaps in the work. The principle applies whether you are joining tubing fittings, sheet metal, or two wire ends twisted together.

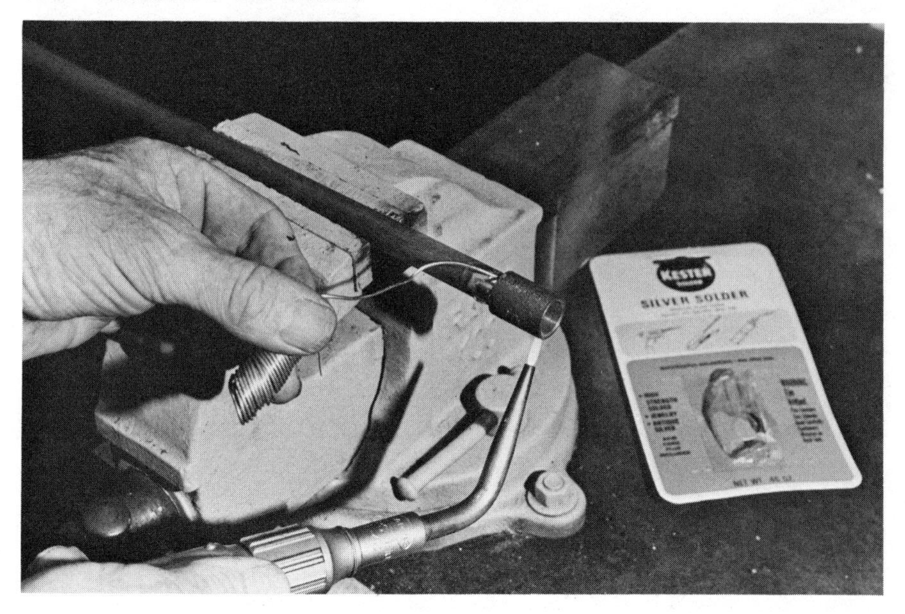

Direct heat from oxyacetylene torch is too hot for silver soldering of copper tube fitting seen here. Note that flame is applied only indirectly by outer flame envelope. At right temperature solder will melt instantly and run into joint by capillary action.

Silver solder. Very often filler metals which melt at a lower temperature than brazing rod but a higher temperature than soft tin/lead solder are classed as silver solder. Most have a light whitish or gray color. There are a great many of these proprietary alloys, some with silver, some with nickel or other metals in combination to accomplish an often very specific performance objective. These special objectives range from duplicating the color of the work metal in jewelry and fine-arts work to resisting corrosion or some unusual agent in refrigeration or chemical-handling pipes.

Unless you take up a specialty hobby you will probably have little reason to use these, but you may find the hardware-store variety of silver solder very useful for some repairs where you need more strength and heat resistance than you can get with soft solder or where you don't want the bronze color or the heat required with brazing. Almost always the nature of the jobs using silver solder requires a very small torch tip and low pressure to the torch. Often the merest touch of the oxyacetylene flame is sufficient to produce a good job. Be very delicate in your handling of such solders, and if a special flux is required follow the directions exactly.

Soft solder. There is seldom any reason to braze copper tubing fittings unless there is an unusual demand for strength in the joint. Usually tin/

lead soft solder will do as well. And the use of oxyacetylene for soft solder is wasteful, costly, and often less successful because of excessive heat. For these reasons I strongly recommend the new oxygen/MAPP or propane torches or even high-output MAPP/air or propane/air torches. The chances of ruining a joint are less and the cost is much lower. Follow the instructions that come with these torches.

ALUMINUM WELDING. Although aluminum has been professionally welded with oxyacetylene for many years I do not recommend doing it other than as an experiment in the home shop. TIG welding will generally prove much more successful. Although I have discussed them previously in the chapter on the carbon-arc torch, I'll repeat the reasons for difficulty in welding aluminum:

- Most aluminum alloys melt slightly below 1300° F. without showing the "range" typical of steel between mushy soft and totally fluid.
- There is no distinct color change to warn you that you're near melting temperature.
- Aluminum forms an oxide on the surface almost instantly, and this must be penetrated or removed to weld.
- Hot aluminum is very weak, and the weld area must often be supported or backed up.
- Aluminum expands much more than steel and tends to distort.
- There are many alloys and hardness grades of aluminum which complicate welding and in some cases make welding almost impossible.
- Thorough cleaning and some form of flux is needed.

If you wish to experiment, obtain some soft aluminum sheets about 1/16" thick or even thicker. Carefully clean them with trisodium phosphate or other alkaline washing agent. Mix aluminum welding flux powder, 3 parts flux to 1 part water. Secure the plates in a lap-weld position and apply the flux paste with a clean brush. The torch tip should be the same size as for steel, since in spite of its low melting point aluminum conducts heat away very rapidly. Do not back up the weld area with a steel plate, but for support try placing a sheet of asbestos paper between the aluminum and the plate. Adjust the torch for a definite feather with slightly excess acetylene and hold the torch at a shallow angle of about 30°. You should be able to see the puddle start in time to move the torch, but don't be discouraged if you burn through many, many times before learning the trick.

Try to run a bead on the lap weld without adding filler. If you manage to do this after practice, try it with aluminum filler rod dipped in the flux.

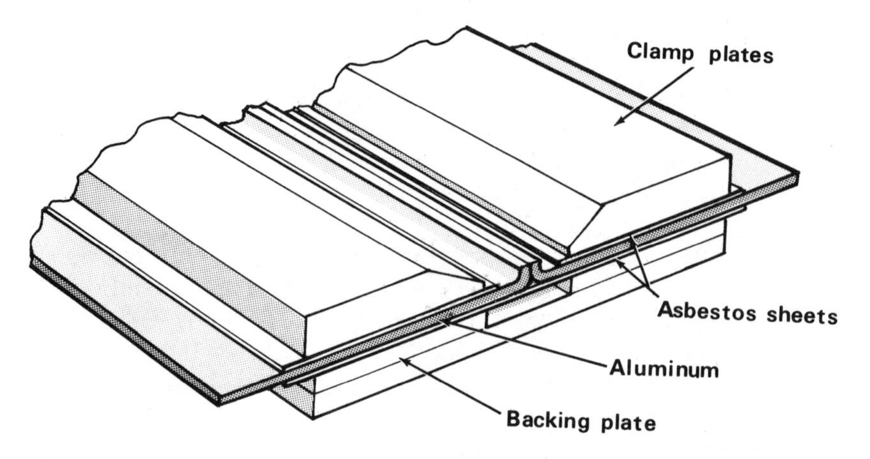

Best way to develop touch for gas welding of aluminum is to start out with an ideal set-up such as upturned edge welds supported on a backing plate. Asbestos allows for slight movement as metal expands. Few actual welds are this nicely arranged, but in the beginning it's more important to learn how aluminum behaves and to recognize when it's ready to fuse.

The techniques and movements do not differ greatly from steel but a very high degree of alertness is required.

Although lap welds, butt welds, and tee welds can be made the same as with steel, many aluminum joints are best designed for edge welding where two bent-up flanges are brought together. Clean and flux thoroughly. Such welds can be made without filler rod simply by melting the two edges together.

Finally, I do not suggest trying an aluminum weld on anything you don't want to risk destroying. Castings are equally tricky. Remember that part of being a successful home-shop weldor is knowing when to turn to the pro.

Annealing aluminum. If you have ever admired the streamlined wheel pants, cowlings, and fairings on antique airplanes, you should know that many of these were hand-formed with no stamping dies at all. The technique can be used on the readily available hardware-store aluminum to produce some unusual projects, or for home-built aircraft. It requires annealing aluminum to a dead-soft state. To anneal aluminum, especially the extruded bars and shapes from the hardware store, it is only necessary to heat it to a critical temperature very evenly and let it cool. Here's how to do it.

Light the torch, using straight acetylene, no oxygen, and a small flame. Play the smoky flame rapidly over the metal and deposit an even coat of carbon. Now open the torch oxygen valve and adjust for a neutral flame. You'll find that moving the flame over the carbon-coated surface will cause the carbon to disappear before your eyes and restore the bright appear-

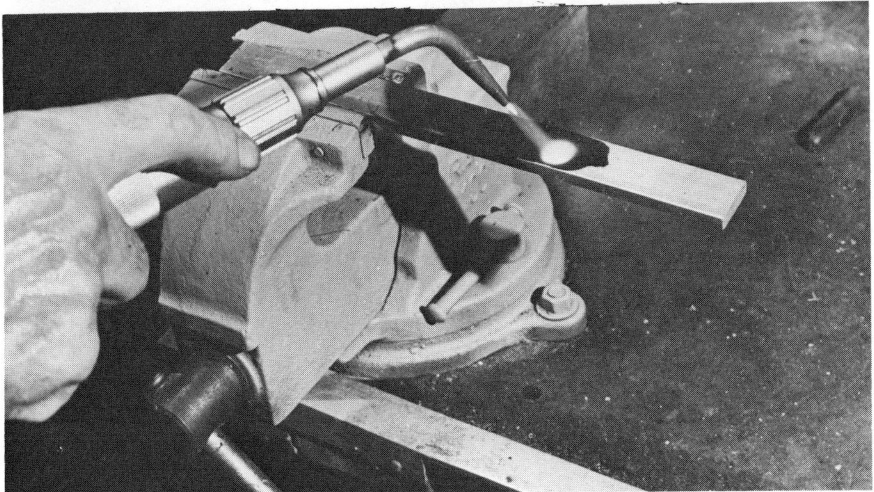

To anneal aluminum, first coat it with carbon from the acetylene flame, top. Carbon serves no purpose except to show you when entire area has been brought to proper temperature in the next step, below, which uses neutral flame to burn off carbon. When all of carbon is gone, metal is soft.

ance. When all of the carbon has been burned away, the aluminum is annealed and is now very soft so it can be formed readily. You can shape it by bending, hammering with a soft-face hammer and a sand bag support, rolling, or any other method. As you work, you'll discover that the metal once again becomes hard and springy. This is called "work hardening." The same hardening will occur automatically in a few days even though you don't work the metal.

The old-time aircraft parts were made by annealing, forming, and annealing again, over and over. The metal was fairly thick to start with to allow hammering and to permit a final dressing to remove tool marks. If you have a job requiring aluminum rivets, the same technique can be used to soften them for heading. To keep them soft for several days, store them in a freezer.

18 Oxyacetylene Cutting

Learning to cut with the oxyacetylene torch is fun and makes otherwise almost impossible jobs easy. I place the topic at the end of this book much as a cook serves dessert after a heavy meal, because I think you'll enjoy it.

Many users of oxyacetylene outfits use their rigs almost exclusively for cutting and do most welding with the arc. Once you've learned the art of flame cutting—and it really is an art—you'll wonder how you ever did without it. Flame cutting makes projects possible which you wouldn't attempt otherwise simply because there's no other practical way to cut heavy metal to the sizes and shapes needed.

In the case of some repairs the oxyacetylene flame is the only way to remove damaged sections. An example might be a boat trailer or a pier structure damaged in an accident. It's true that you can cut with the electric arc, but it's difficult to make clean-edged, intricate, and accurate cuts such as you can make with the flame. And don't forget, some of the little portable outfits that use MAPP gas or propane together with a small oxygen bottle will do a fine job of cutting.

WHAT FLAME CUTTING IS. In earlier chapters we discussed the construction of cutting torches and described how the nozzles have four or six orifices surrounding a central orifice. Review this section to get a better understanding of your cutting torch. Summed up, the outside orifices are adjusted to a neutral flame, but with a slightly different procedure, as I'll explain. These preheat orifices bring a spot on the work to a red, or kindling, temperature. At this point the oxygen cutting lever is depressed and a stream of fairly high-pressure oxygen emits from the central orifice and causes the heated metal to burn rapidly. Once started, the process tends to feed on itself and the cut may be guided as needed.

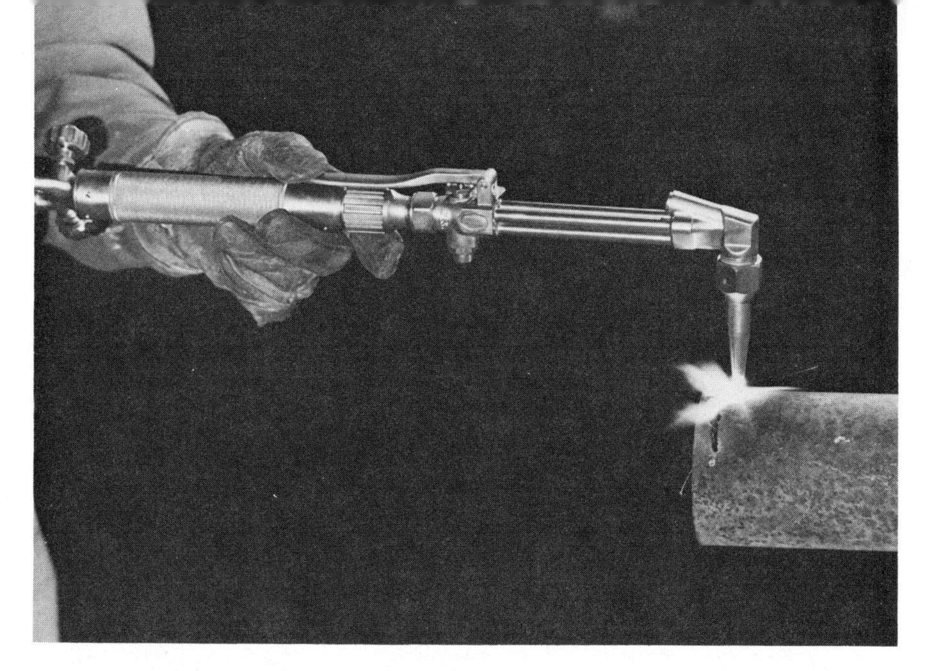

Although this picture shows cutting torch in action with cutting lever depressed, it also shows about the worst way to hold the torch. With this grip it's almost impossible to hold it steady or maintain flame clearance.

Miniature of large flame-cutting machine is motorized torch with small wheels to maintain flame clearance and an adjustable-speed motor drive to move torch at a constant rate. Even so, careful flame adjustment is needed. As you practice, try to train your hand to duplicate as closely as possible the constant, precision movement of such machines.

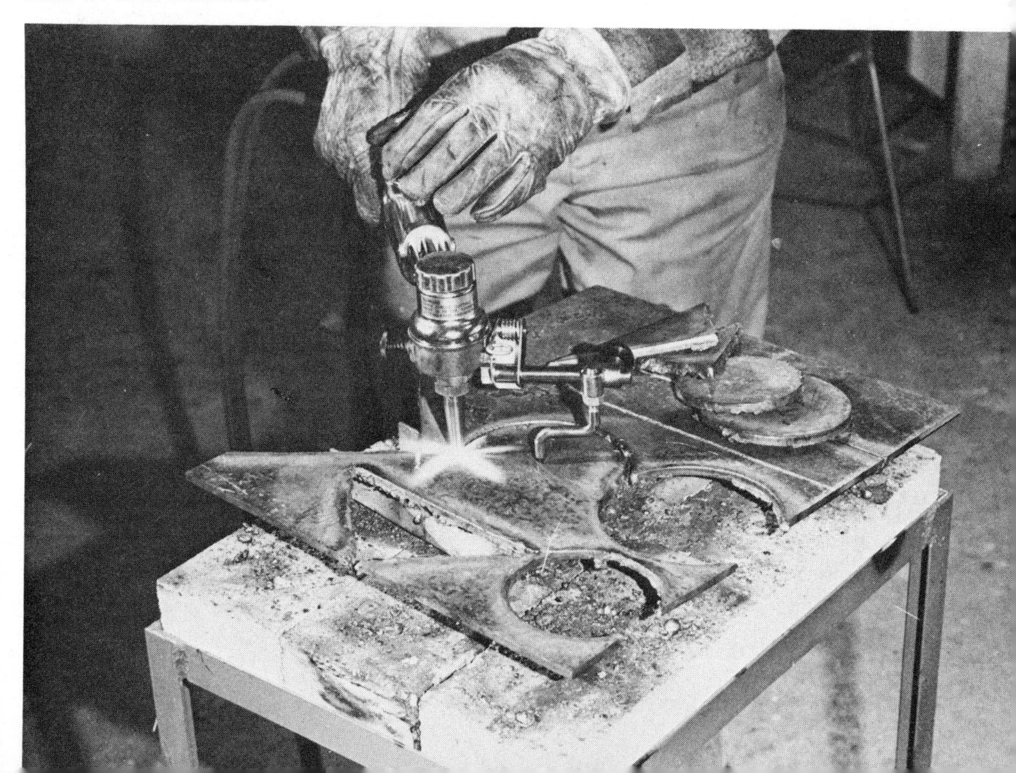

But guiding the hand-held torch, keeping it distanced accurately from the work surface and moving at just the right rate, and recognizing whether you have too much or too little heat add up to quite an art. If you've ever seen an oxyacetylene cutting machine turning out heavy pieces, often an inch or more thick, with a fine line cut and an almost smooth edge, you know that such cuts are possible. They are almost equally possible with your hand-held torch, but in the beginning expect to get some messy, jagged, miscut practice pieces.

ADJUSTING THE TORCH. As you gain in experience you'll learn to recognize the small differences in cutting performance resulting from torch adjustment. But to get started, it's wise to stick to the pressures and nozzle sizes recommended by the maker of your equipment. Note that the nozzle sizes in the table refer to only one manufacturer's numbering code. They run from rather small to medium, and you can quickly relate them to your own equipment by comparing the orifice sizes.

Cutting Nozzle Size, Metal Thickness, And Gas Pressures

Metal Thickness (in.)	Nozzle Size	Oxygen Pressure (psi)	Acetylene Pressure (psi)
1/8 to 1/4	00	25–35	6
3/8	0	25–35	6
1/2 to 1	1	35–45	6
2	2	45–50	6

There's nothing hard and fast about such recommendations, and you will want to experiment a bit later. Note the following differences in flame adjustment from what you did in welding:

- You must open the oxygen valve on the torch handle wide open.
- Open the acetylene valve about a quarter-turn and ignite the torch.
- Use the oxygen valve on the cutting attachment to adjust the preheat flames to show a slight acetylene feather at first.
- Press down on the cutting oxygen valve and watch the preheat flames. With this valve still open, use the oxygen valve on the cutting attachment to make a final adjustment for a neutral flame. All orifices should project the same flame length and all flames should have the same appearance.

Correct Procedures vs. Common Faults in Hand-Cutting

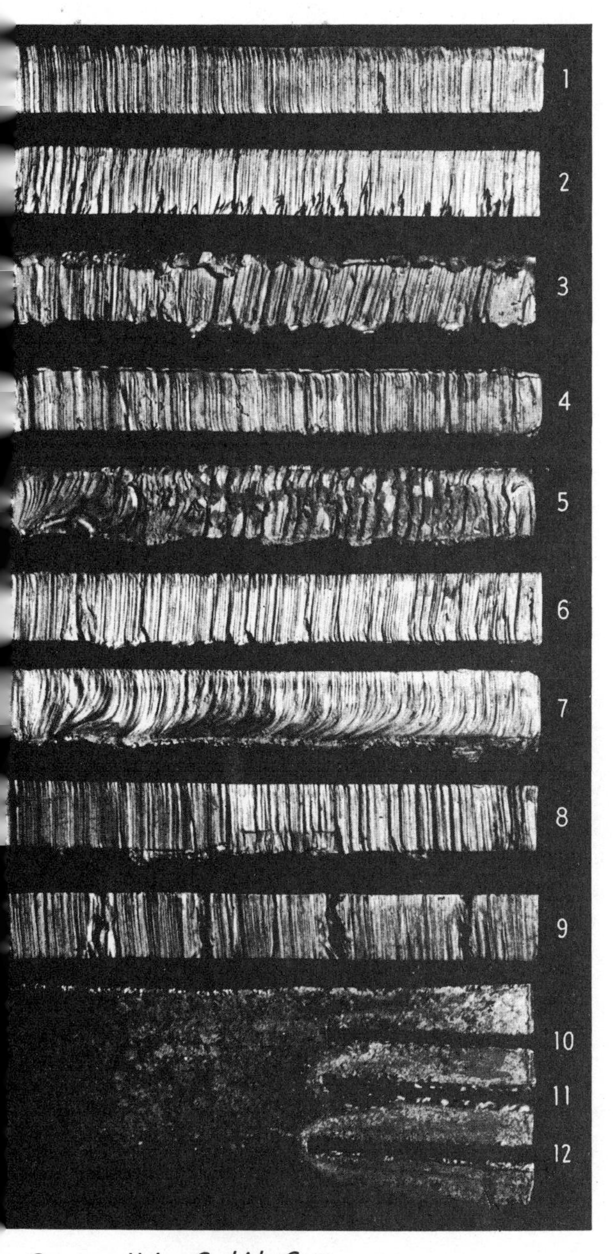

Courtesy Union Carbide Corp.

(1) Correct Procedure

Compare this correctly made cut in 1-in. plate with those shown below. The edge is square and the drag lines are vertical and not too pronounced.

(2) Preheat Flames Too Small

Fault: preheat flames too small—only about $\frac{1}{8}$ in. long. Result: cutting speed was too slow, causing bad gouging effect at bottom.

(3) Preheat Flames Too Long

Fault: preheat flames too long—about $\frac{1}{2}$ inch. Result: top surface has melted over, the cut edge is irregular, and there is too much adhering slag.

(4) Oxygen Pressure Too Low

Fault: oxygen pressure too low. Result: top edge has melted over because of too slow cutting speed.

(5) Oxygen Pressure Too High

Fault: oxygen pressure too high and nozzle size too small. Result: entire control of the cut has been lost.

(6) Cutting Speed Too Slow

Fault: cutting speed too slow. Result: irregularities of drag lines are emphasized.

(7) Cutting Speed Too High

Fault: cutting speed too high. Result: there is a pronounced rake to the drag lines and the cut edge is irregular.

(8) Blowpipe Travel Unsteady.

Fault: blowpipe travel unsteady. Result: the cut edge is wavy and irregular.

(9) Lost Cut Not Properly Restarted

Fault: cut lost and not carefully restarted. Result: bad gouges were caused where cut was restarted.

(10) Good Kerf

Compare this view (from the top of the plate) of a good kerf with those below. This cut was made by using correct procedures.

(11) Too Much Preheat

Fault: too much preheat and nozzle too close to plate. Result: bad melting over the top edge occurred.

(12) Too Little Preheat

Fault: too little preheat and flames too far from plate. Result: heat spread has opened up kerf at top. Kerf is too wide and is tapered.

You may or may not have the correct adjustment. If the preheat flames burn out, away from the nozzle, reduce the acetylene flow with the torch valve. Readjust the oxygen as before. If you don't have enough preheat flame it will take longer to bring the steel to a kindling temperature and the cut will have an eroded, jagged appearance on the bottom. If your preheat flames are too long, you'll see a line of melted metal on the top, the cut will be ragged, and there will be slag hanging from the bottom.

MAKING THE CUT. Once you've learned to adjust the flame, the next step is to learn to move the torch along smoothly and evenly just as close as you can approximate the cutting machine in the factory. In real life your joints and muscles won't work quite that way, but some techniques will help.

At first, if you've already learned to arc-weld or gas-weld, the natural inclination is to hold the cutting torch in one hand the way you might for gas welding. This will work, but very badly. You will not be able to hold the spacing of the nozzle accurately.

Steadier way to hold torch uses one hand as a support or roller, and the other to hold torch and make adjustments to follow line. Action is more easily controlled in this position.

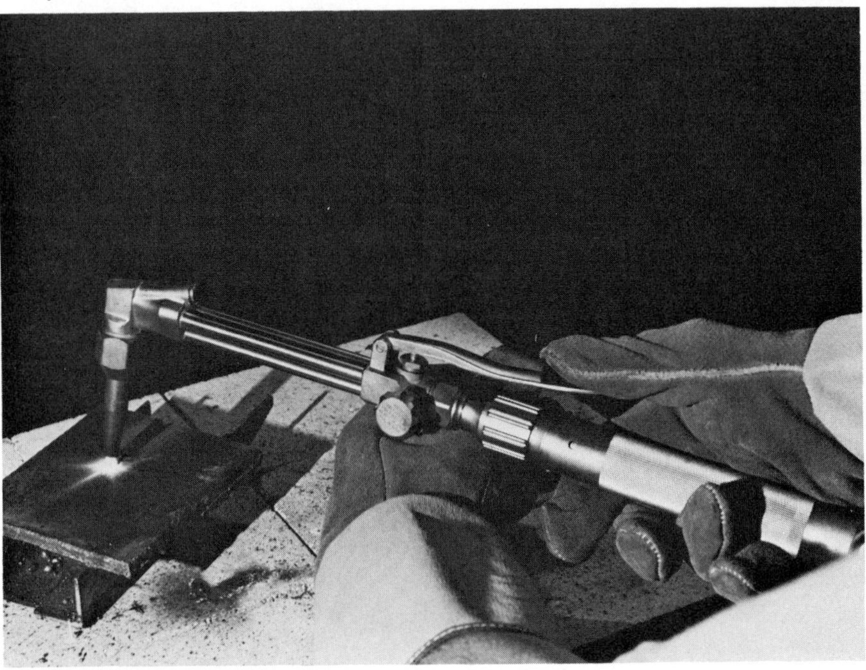

But there's a better way. In fact, almost every experienced flame cutter has his own pet way to move the torch along the desired line and at the same time hold the preheat flames to a constant relationship with the metal. For a beginning, try using your other hand, your left if you're right-handed, as a sort of support and roller or spacer. With your right hand holding the torch and your thumb in place on the cutting oxygen lever, bring your left hand underneath so the knuckles are resting on the work metal. Support the torch on the lightly clenched fingers of your left hand. Now, if you start at the right and slowly roll your left hand along the work surface, you've got sort of a steady rest and height gage. Since your wrist joint is not a perfect mechanical rotating pivot, the rolling action will move the torch in a sweep or arc. Use the right hand to guide the torch and slide it forward or back to follow the line. I suggest that you practice this without the flame until you develop a feel for the somewhat unnatural movement of your left hand.

Start with a piece of ¼″ or ⅜″ scrap steel and position it so the movements above will be comfortable. If you choose a sitting position a leather apron is worth having, since there will be a heavy discharge of slag from the cut. Be certain to have the workpiece elevated from the bench so the slag can clear underneath. Some scrap channel iron with the side flanges turned up will do for such a support. All of the safety practices—close-fitting goggles, full arm and leg protection, and sturdy shoes with no openings for metal to enter—are extremely important in flame cutting. So, of course, is fire prevention, and you should expect to see a shower of sparks and slag. Some you might catch in a pan or bucket of sand, but be assured that some little white-hot globules will go bouncing off into the nether regions of your shop.

Now light and adjust the torch and position your hands as described. The flames should be almost touching the right edge of your workpiece. Do nothing with the cutting oxygen valve for the moment. Move the torch so the nozzle is vertical to the work and all flames are even with respect to the metal surface. The typical bright-red glow will develop very quickly, and when the steel just starts to melt, slowly squeeze the oxygen cutting lever and at the same time re-aim the torch nozzle just a bit in the direction you intend to cut.

If the metal has been properly preheated a discharge of molten slag will emit instantly from under the work. If this doesn't happen, release the oxygen lever and heat for a few seconds more. You'll soon learn the right moment to introduce the cutting oxygen.

Holding the line. Once the cut has started there must be no hesitation. Remember, you're trying to duplicate the smooth, steady motion of a cutting machine. You may be surprised at first at just how rapid the cutting action is. Move the torch right along. If you lose the cut, merely repeat the starting procedure from where you lost it and go on, this time a trifle slower.

The limitations of human joints will prevent machinelike perfection. You can move only a limited distance along the line, perhaps 3″ or 4″, before you have to reposition your hands. Practice straight cuts along ruled lines until you can make them fairly well, then lay out some circular cuts of various diameters. You'll find it helps if you can place the circle tangent to an edge at some point to give you a starting point on an edge.

We are not all constructed alike. Some have hands that cannot span an octave on a piano. Some have past injuries or arthritis that inhibit the smooth, free-rolling action of the hands that I've described. There's nothing wrong with giving yourself a little help if you find it extremely difficult to get a smooth, even progression of the cutting torch. One thing that's worth trying is placing a block of wood or metal back 8″ or 10″ from the cut to elevate your hand and wrist. The approach is somewhat like a sign painter with a stick supporting and steadying his wrist while he paints in letters. I recall one expert professional iron worker who used a cutting torch all day long and always carried a fat 2″-diameter roll of asbestos

You'll probably cut up a lot of scrap metal before you acquire the ability to make a clean cut right on the line every time. Jagged edges, slag along the bottom, and crooked cuts are all part of learning

paper bound with wire in his pocket. He'd sit high in the air on a steel beam and use this roll as a sort of pillow for his wrist while he trimmed heavy structural pieces with a surgeon's accuracy.

If you want to make a true straight-line cut, there's no law that you can't clamp a piece of angle iron or the like so you can run the torch handle against it much as you'd rule a line with a pencil. Part of the fun of welding is using your own wits to solve problems. Just because a pro who does the same thing day after day can carry off a job freehand doesn't mean that you have to do it on a once-in-a-lifetime job.

MAKING HOLES. The technique for burning holes, called piercing, is about the same as for a cut, with the exception that you'll be holding the torch in one spot. Since you're not moving along a line you should keep the torch vertical to the surface. There's nothing hard about that except that you also want to avoid having gobs of slag blow out of the molten puddle and plaster over the holes in the torch nozzle. This can cost the price of a new nozzle. There are three tricks to avoid this—most of the time.

- When the molten puddle appears, open the oxygen lever very slowly.
- At the same time raise the torch from the metal to ½″ to ⅝″ clear of the plate.
- Simultaneously, move the torch in a slight circular pattern.

Once the hole has been punched through, you should continue to work around the edges in a circular pattern until you get the size you want. For metal of the thickness commonly used in the home shop, none of this is difficult. Obviously, it becomes more difficult if you are piercing 1″ or 2″ plate with more molten slag to clear from the bottom.

WHAT IS A GOOD CUT? At first you'll probably settle for almost any cut that haggles off the piece you want. But this doesn't produce shapes that can be fitted and welded together neatly. And you have the choice of extensive grinding and cleaning up or producing a finished job that makes strong men shudder to view.

So, a good cutting job will have clean edges and follow the layout line accurately. Some prefer to cut right down the line. Others find it easier to cut just to the outside. If you're unsure of yourself, cut slightly oversize and grind or power-sand to final dimension if it's critical. To improve your cutting, compare the cut edges of your work with the accompanying pictures. Meanwhile, enjoy your welding and cutting skills.

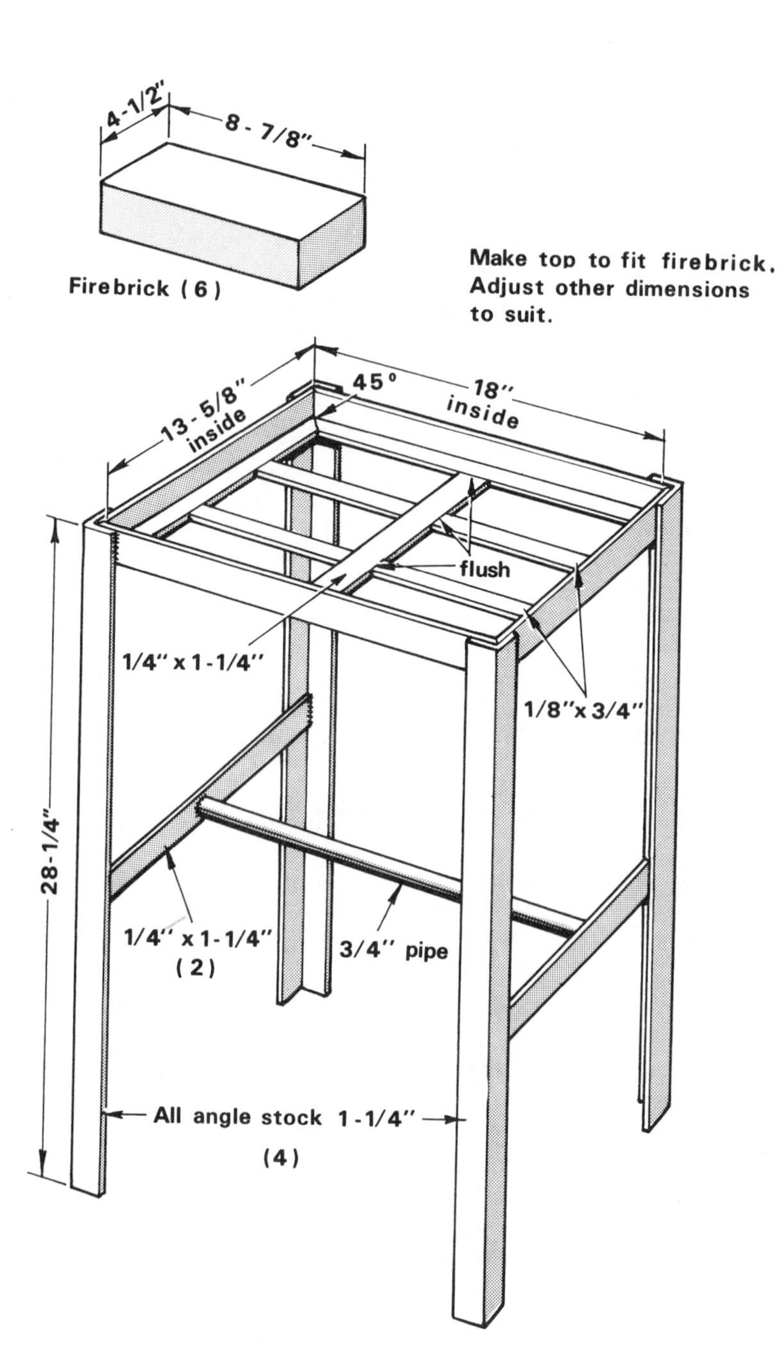

Welding bench with firebrick top is large enough for most home jobs. Dimensions of angle-iron top must be adapted to *your* firebrick. They aren't all alike. Put bricks loosely together and cut and weld top around them. Don't fit them too snugly but leave clearance so you can lift them out. Weld with gas or arc. Bench can be welded with only a 50-amp welder, but 70 to 100 amps is better for this size iron.

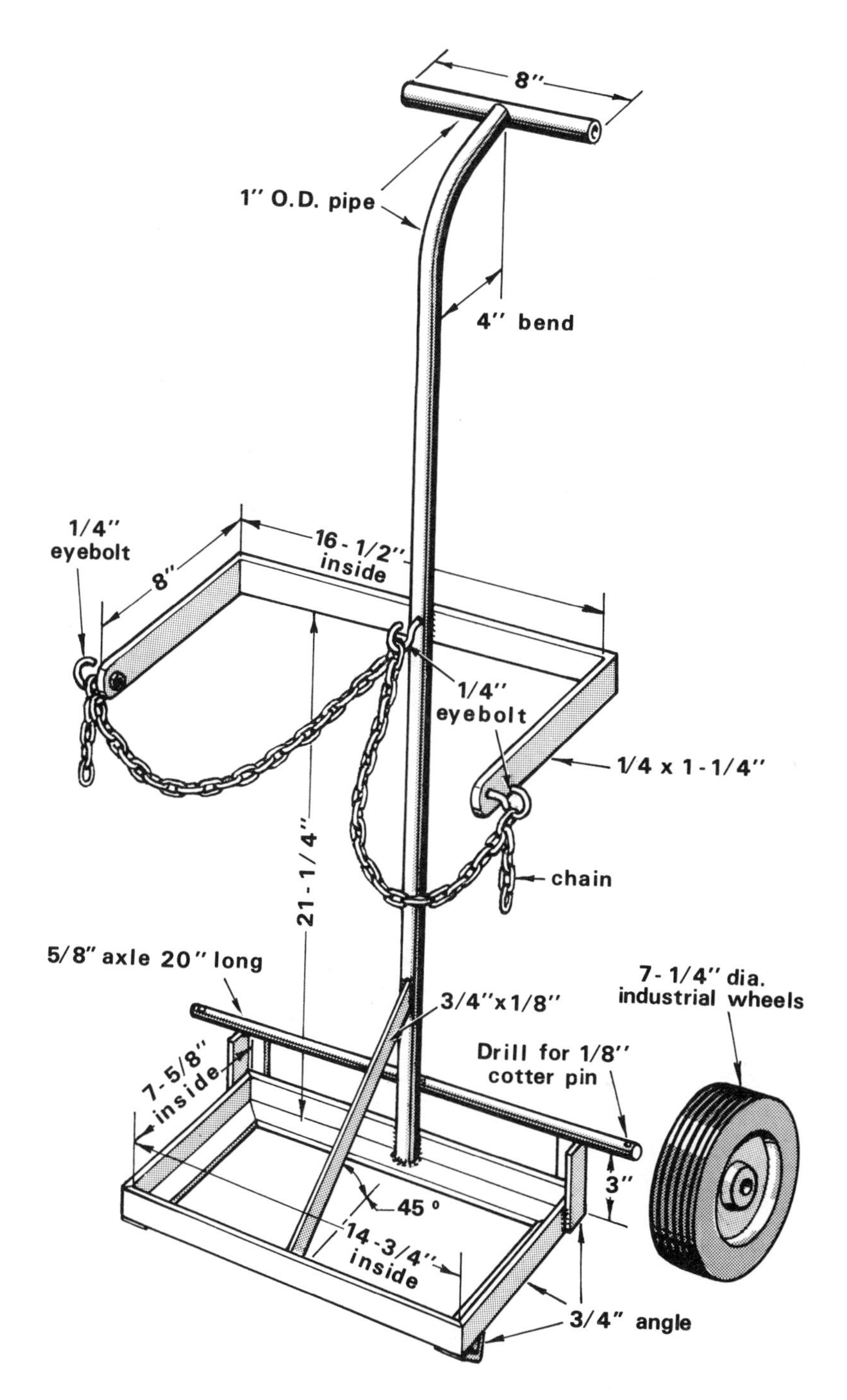

Dimensions of tank cart can vary slightly. Be careful of two things. First, fit bottom support to *your* tanks so there will be room for chain between them, and so gages and regulators won't jostle each other. Second, axle location depends on wheel diameter. Make cart and add small feet before welding axle in place. When cart is upright, wheels should just touch floor but carry no weight. Locate and weld axle with wheels in place and cart resting on feet on level floor.

Index